D1295726

ISBN-13: 978-0-13-784190-5
ISBN-10:   0-13-784190-6

9 780137 841905

90000

Electronics Pocket Handbook

Third Edition

Daniel L. Metzger

Monroe County Community College

To join a Prentice Hall PTR Internet mailing list, point to:
http://www.prenhall.com/mail.lists/

Prentice Hall PTR
Upper Saddle River, New Jersey 07458
http://www.phptr.com

Library of Congress Cataloging-in-Publication Data

Metzger, Daniel L.
   Electronics pocket handbook / Daniel L. Metzger, -- 3rd ed.

     p. cm.
   Includes index.
  ISBN-10: 0-13-784190-6
  ISBN-13: 978-0-13-784190-5
    1. Electronics -- Handbooks, manuals, etc.   I. Title.
TK7825.M45 1998
621.381 '0212 -- dc20   91-34052
                                        CIP

Editorial / production supervisor: *Craig Little*
Manufacturing manager: *Julia Meehan*
Acquisitions editor: *Bernard Goodwin*
Cover design director: *Jerry Votta*
Cover designer: *Karen Marsilio*

© 1998 Prentice Hall PTR
Prentice-Hall, Inc.
Upper Saddle River, New Jersey 07458

The publisher offers discounts on this book when ordered in bulk
quantities. For more information, contact Corporate Sales Department,
Phone: 800-382-3419, Fax: 201-236-7141; email:
corpsales@prenhall.com or write Corporate Sales, Prentice Hall PTR,
1 Lake Street, Upper Saddle River, NJ 07458

Prentice Hall books are widely used by corporations and government
agencies for training, marketing, and resale.

Printed in the United States of America

13, 07/07

ISBN-10: 0-13-784190-6
ISBN-13: 978-0-13-784190-5

Prentice-Hall International (UK) Limited, *London*
Prentice-Hall of Australia Pty, Limited, *Sydney*
Prentice-Hall Canada Inc., *Toronto*
Prentice-Hall Hispanoamericana, S.A., *Mexico*
Prentice-Hall of India Private Limited, *New Delhi*
Prentice-Hall of Japan, Inc., *Tokyo*
Editora Prentice-Hall do Brasil, Ltda., *Rio de Janeiro*

Contents

1

Definitions, Formulas, and Charts

1.1 DEFINITIONS OF ELECTRICAL QUANTITIES

Charge. Quantity symbol: Q, implying *quantity* of electricity. Unit symbol: C (Coulomb); the charge possessed by 6.242×10^{18} electrons, or the same number of protons. The basic electrical quantity, although seldom measured for lack of a suitable instrument.

Current. Quantity symbol: I, from the French *intensite.* Unit symbol: A (Ampere). A flow of charge equivalent to 1 Coulomb per second. Commonly measured by breaking the circuit path and forcing the current to flow through the ammeter.

Voltage, also called *potential, potential difference*, and *electromotive force (EMF.)* Quantity symbols: E and V both acceptable. Some writers use E for voltage sources and V for voltage drops. Unit symbol: V (Volt). The potential for pushing current between two points, if a conductive path is provided. The most commonly measured electrical quantity, because it involves simply touching the voltmeter probes to the two points.

Resistance. Quantity symbol: R. Unit symbol: Ω (Ohm). A ratio of voltage divided by current for a given conductor under a given set of conditions. Commonly measured by disconnecting one end of the device from any other circuit connections, then applying a voltage and measuring the resulting current, or applying a current and measuring the resulting voltage.

Conductance. Quantity symbol: G. Unit symbol: S (Siemens.) The reciprocal of resistance: $G = 1/R$. Less frequently measured than resistance.

Energy. Quantity symbol: W, implying *work*. Unit symbol: J (Joule.) The work done when a voltage of 1 V pushes a current of 1 A through a resistance for a time of 1 second. Equal to the work done when a force of 1 newton acts through a distance of 1 meter. Seldom measured in the laboratory, but commonly recorded by utility companies with *watt-hour* meters.

Power. Quantity symbol: P. Unit symbol: W (Watt.) A rate of doing work equal to 1 joule per second. The rate of energy expenditure when a potential of 1 volt causes a current of 1 ampere. Measurable by wattmeters, which indicate the product of voltage and current measurements.

1.2 OHM'S LAW AND POWER LAW

The commonly measured electrical quantities are voltage, current, resistance, and power. If any two of these are known, the other two can be calculated:

$$I = \frac{V}{R} = \frac{P}{V} = \sqrt{\frac{P}{R}}$$

$$V = IR = \frac{P}{I} = \sqrt{PR}$$

$$R = \frac{V}{I} = \frac{P}{I^2} = \frac{V^2}{P}$$

$$P = IV = I^2R = \frac{V^2}{R}$$

In the memory aids below, cover up the desired quantity to see the formula for calculating it:

1.3 RESISTOR COMBINATIONS

Series resistors may be replaced by a single resistor equal to their sum. Resistors are in series if the identical current flows in each of them, regardless of whether they are actually connected end-to-end. Resistors are *not* in series if any third element that may carry current is connected to their junction.

$$R_S = R_1 + R_2 + R_3 + \cdots$$

Parallel resistances. Resistors are in parallel if they are connected together at one end, and also connected together at the other end, regardless of whether they actually appear side-by-side. Any number of parallel resistances may be converted to conductance units and reduced to a single resistance having a conductance equal to the sum of all parallel conductances:

$$G_P = G_1 + G_2 + G_3 + \cdots$$

Without using conductance units, this procedure becomes:

$$R_P = \cfrac{1}{\cfrac{1}{R_1} + \cfrac{1}{R_2} + \cfrac{1}{R_3} + \cdots}$$

With a calculator, proceed as follows:

$$R_1 \boxed{1/x} \boxed{+} R_2 \boxed{1/x} \boxed{+} R_3 \boxed{1/x} \boxed{=} \boxed{1/x}$$

For two resistors in parallel, use the "product over the sum." *This formula cannot be extended for three resistors.*

$$R_P = \frac{R_1 \times R_2}{R_1 + R_2}$$

Parallel design. To find R_2 such that R_2 in parallel with R_1 will produce a desired R_P:

$$R_2 = \frac{R_1 \times R_P}{R_1 - R_P}$$

3

1.4 CAPACITORS AND INDUCTORS

A **capacitor** is a two-terminal element that opposes any change in voltage across it by passing a current proportional to the rate of change of voltage. It is impossible to change the voltage across a capacitor instantly, as this would produce an infinite current.

Capacitor Formulas

Charge relationship	$Q = CV$
Reaction against voltage change	$I = C \dfrac{\Delta V}{\Delta t}$
Constant-I charge or discharge	$CV = It$
Time constant	$\tau = RC$
Energy stored	$W = 1/2\, CV^2$
Series combinations	$C_S = \dfrac{1}{\dfrac{1}{C_1} + \dfrac{1}{C_2} + \cdots}$
Parallel combination	$C_P = C_1 + C_2 + \cdots$

An **inductor** is a two-terminal device that reacts against any change in current through it by generating a voltage in opposition to the applied voltage and proportional to the rate of change of current. It is impossible to change the current through an inductor instantly, as any such attempt would produce an infinite voltage.

Inductor Formulas

Charge relationship	$Q = It$
Reaction against current change	$V = L \dfrac{\Delta I}{\Delta t}$
Constant-V charge or discharge	$LI = Vt$

Time constant	$\tau = \dfrac{L}{R}$
Energy stored	$W = \frac{1}{2}LI^2$
Series combinations (no mutual inductance)	$L_S = L_1 + L_2 + \cdots$
Parallel combination	$L_P = \dfrac{1}{\dfrac{1}{L_1} + \dfrac{1}{L_2} + \cdots}$

Mutual inductance is possessed by two coils whose magnetic fields are coupled. Measure L_{AID} with the two coils in series-aiding, and L_{OPP} with them in series-opposing.

$$M = \frac{L_{AID} - L_{OPP}}{4}$$

Where M is known, two inductors in series have total

$$L_T = L_1 + L_2 + 2M \quad \text{(aiding fields)}$$

$$L_T = L_1 + L_2 - 2M \quad \text{(opposing fields)}$$

The coefficient of coupling of two coils is

$$k = \frac{M}{\sqrt{L_1 L_2}}$$

Time constant (τ) is the time required for a process to go 63.2% of the way from where it is to completion, following a *negative-exponential* rising curve given by

$$y = 1 - e^{-t/\tau}$$

or on the falling curve by

$$y = e^{-t/\tau}$$

where y represents the fraction of maximum voltage or current. The graph of Figure 1.1 on the next page shows the curves. Notice that the initial charge or discharge rate would, if continued, bring the process to completion in one time constant, τ.

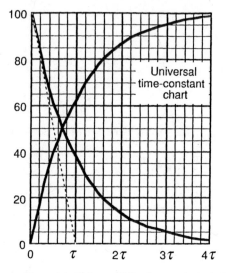

Figure 1.1 Rising and falling time constants.

The following table gives the fractions of full charge reached and remaining to go for various times. The bold-faced lines are often committed to memory.

$\frac{t}{\tau}$	y	$1 - y$
0.1	0.095	0.905
0.2	0.181	0.819
0.3	0.259	0.741
0.5	0.393	0.607
1.0	**0.632**	**0.368**
2.0	0.865	0.135
3.0	**0.950**	**0.050**
4.0	0.982	0.018
5.0	**0.993**	**0.007**
7.0	0.999	0.001

Voltage and current waveforms for a capacitor charging and discharging, when the charging times are much longer than one time constant, are given in Figure 1.2, below.

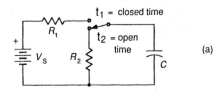

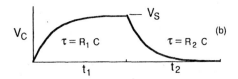

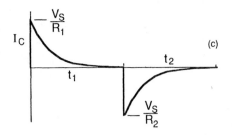

Figure 1.2 Capacitive charging time much longer than τ. Circuit (a), with capacitor voltage (b) and capacitor current (c), charging and discharging.

Fast-switching inputs (which allow much less than one
time constant for charge and discharge) produce waveforms
like the ones in Figure 1.3, below, for an *R-C* circuit.

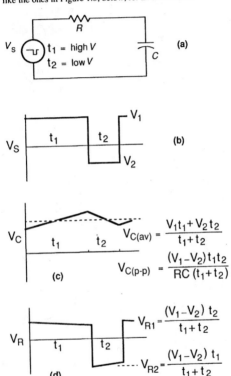

$$V_{C(av)} = \frac{V_1 t_1 + V_2 t_2}{t_1 + t_2}$$

$$V_{C(p-p)} = \frac{(V_1 - V_2) t_1 t_2}{RC (t_1 + t_2)}$$

$$V_{R1} = \frac{(V_1 - V_2) t_2}{t_1 + t_2}$$

$$V_{R2} = \frac{(V_1 - V_2) t_1}{t_1 + t_2}$$

Figure 1.3 Capacitive charging time much shorter
than τ. Circuit (a), source voltage (b), capacitor
voltage (c), and resistor voltage (d).

Inductor voltage and current waveforms for charging and discharging, when the charging times are much longer than one time constant, are given in Figure 1.4, below.

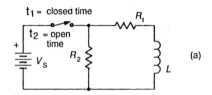

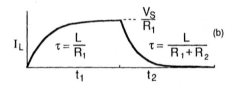

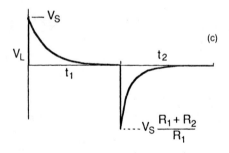

Figure 1.4 Inductive charging time much longer than time constant τ. Circuit (a), inductor current (b), and inductor voltage (c).

Fast-switching R-L circuits (with time constants much longer than the charging and discharging times allowed) produce waveforms like the ones in Figure 1.5, below.

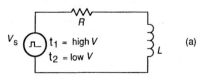

(a)

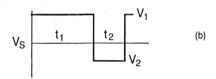

(b)

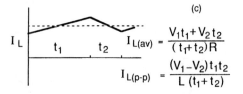

(c)

$$I_{L(av)} = \frac{V_1 t_1 + V_2 t_2}{(t_1 + t_2)R}$$

$$I_{L(p-p)} = \frac{(V_1 - V_2) t_1 t_2}{L (t_1 + t_2)}$$

(d)

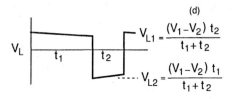

$$V_{L1} = \frac{(V_1 - V_2) t_2}{t_1 + t_2}$$

$$V_{L2} = \frac{(V_1 - V_2) t_1}{t_1 + t_2}$$

Figure 1.5 Inductive charging time much shorter than τ. Circuit (a), source voltage (b), inductor current (c), and inductor voltage (d).

1.5 AC , HARMONICS, RMS, AND SKIN EFFECT

Period T vs. frequency : $\qquad T = \dfrac{1}{f} \qquad f = \dfrac{1}{T}$

Radian frequency ω : $\qquad \omega = 2\pi f \qquad f = \dfrac{\omega}{2\pi}$

Fourier's theorem states that a sine wave is the only pure frequency, and that any distortion of the sine shape represents *harmonic* frequencies present in addition to the fundamental sine wave. Figure 1.6, below, shows that a second harmonic added to the fundamental can approach a half-wave-rectified shape. If more even harmonics (4th, 6th, etc.) are added in correct amplitudes, an almost perfect half-wave can be produced. Figure 1.7 shows that a fundamental and a third harmonic can approach a square wave. A perfect square wave contains an infinite number of odd-numbered harmonics of decreasing amplitudes.

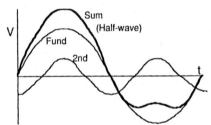

Figure 1.6 Fundamental frequency plus second harmonic approaches a half-wave-rectified waveform.

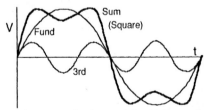

Figure 1.7 Fundamental plus third harmonic.

The **Fourier spectra** for two waveforms are illustrated in Figure 1.8, and the table below lists the harmonic contents of several other common waveforms.

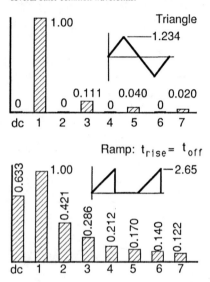

Figure 1.8 Harmonic contents of two waveforms.

Fourier spectra of common waveforms *

Waveform	peak	dc	2nd	3rd	4th	5th	6th	7th
Square	0.785	0	0	0.33	0	0.20	0	0.14
Sawtooth	1.571	0	0.50	0.33	0.25	0.20	0.17	0.14
Half-wave rectified	2.00	0.64	0.42	0	0.08	0	0.04	0
Full-wave rectified	2.356	1.5	0.20	0.09	0.05	0.01	0.01	0.01

* Fundamental-frequency amplitude = 1.00.

When to use *peak, average,* and *rms*:

1. Use peak voltage or current to calculate maximum *instantaneous* power only. Use V^2/R or I^2R.

2. Use average current to calculate average power when the voltage is fixed dc. Use average voltage to calculate average power when the current is unvarying dc.

3. Use rms voltage and/or rms current to calculate average power when the load is a linear device and both V and I are ac in phase. Use IV, I^2R, or V^2/R.

4. RMS measure is assumed in any ac voltage or current notation unless peak, peak-to-peak, or average is specified.

5. The factor 0.707 for converting peak to rms applies to sine waves only. The figures below and on the next page give the factors for other common waveshapes.

6. VOMs and most DVMs are calibrated to read $1.11 V_{avg}$ of an ac wave. VTVMs and FET meters generally read $0.707 V_{pk}$. For nonsinusoidal waveforms the resultant reading is *not* V_{rms}.

The rms value of a nonsinusoidal wave V_t can be obtained from the rms values of its harmonic components V_n by:

$$V_t = \sqrt{V_1^2 + V_2^2 + \cdots + V_n^2}$$

Figure 1.9, below and on the following pages, relates peak, rms, and average values for common waveforms.

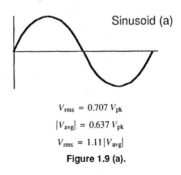

Sinusoid (a)

$$V_{rms} = 0.707\, V_{pk}$$
$$|V_{avg}| = 0.637\, V_{pk}$$
$$V_{rms} = 1.11\, |V_{avg}|$$

Figure 1.9 (a).

Symmetrical trapezoid (b)

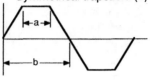

$$V_{rms} = \frac{a + 0.577\,(b - a)}{b}\,V_{pk}$$

$$|V_{avg}| = \frac{a + b}{2b}\,V_{pk}$$

DC pulse (c)

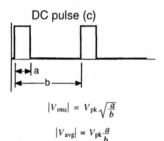

$$|V_{rms}| = V_{pk}\sqrt{\frac{a}{b}}$$

$$|V_{avg}| = V_{pk}\frac{a}{b}$$

Triangle or sawtooth (d)

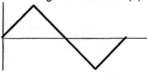

$$V_{rms} = 0.577\,V_{pk}$$

$$|V_{avg}| = 0.500\,V_{pk}$$

Figure 1.9 (b), (c), and (d).

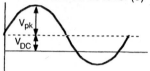

Sine wave on dc level (e)

$$V_{rms} = \sqrt{V_{DC}^2 + \frac{V_{pk}^2}{2}}$$

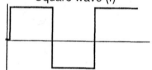

Square wave (f)

$$V_{rms} = V_{pk}$$

$$|V_{avg}| = V_{pk}$$

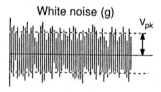

White noise (g)

$$|V_{rms}| \approx \frac{1}{4} V_{pk}$$

Figure 1.9 (e), (f), and (g).

Figure 1.10 and the table below give the rms and average values of the output of an SCR and of a triac for various conduction angles, ϕ. The SCR delivers output only on the positive half-cycle, as shown by the upper shaded area. The triac delivers an additional negative output with an equal conduction angle, represented by both shaded areas.

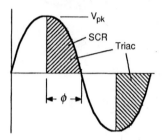

Figure 1.10 SCR and triac outputs

ϕ	Half-wave (SCR)		Full-wave (triac)	
	$\dfrac{V_{rms}}{V_{pk}}$	$\dfrac{|V_{avg}|}{V_{pk}}$	$\dfrac{V_{rms}}{V_{pk}}$	$\dfrac{|V_{avg}|}{V_{pk}}$
15°	0.028	0.008	0.040	0.016
30°	0.085	0.022	0.120	0.044
45°	0.15	0.045	0.21	0.090
60°	0.22	0.078	0.31	0.156
75°	0.29	0.115	0.41	0.230
90°	0.35	0.16	0.49	0.32
105°	0.41	0.21	0.58	0.42
120°	0.45	0.24	0.64	0.48
135°	0.48	0.27	0.68	0.54
150°	0.49	0.29	0.69	0.58
165°	0.50	0.31	0.71	0.62
180°	0.50	0.32	0.71	0.64

Skin effect increases the resistance of wire at high frequencies. For a copper wire in free space the increase becomes serious when the skin depth δ becomes less than one-half the diameter of the wire, as shown in Figure 1.11.

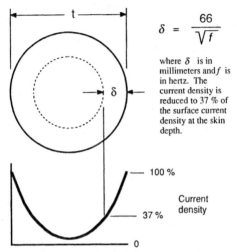

$$\delta = \frac{66}{\sqrt{f}}$$

where δ is in millimeters and f is in hertz. The current density is reduced to 37 % of the surface current density at the skin depth.

Figure 1.11 Skin depth vs. current density.

The resistance of a copper wire in free space under skin effect is

$$R = 8.3 \times 10^{-5}\, d\, \frac{\sqrt{f}}{t}$$

where d is length in meters, t is diameter in millimeters, and f is in hertz. For other metals, multiply R by the square root of the resistivity relative to copper.

Wires in bundles or wound in coils suffer from skin effect at substantially lower frequencies because of *proximity effect*. The resistance given by the foregoing equation should be multiplied by approximately 10 for single-layer coils, and by 50 to 100 for multilayer coils.

Wire sizes (AWG) having resistance increase from skin effect of approximately 10 %. Larger wire sizes suffer losses above 10 %.

f	Single straight wire	Single-layer coil	Multilayer coil
60 Hz	—	0	8
400 Hz	00	4	11
1 kHz	3	6	14
10 kHz	13	14	27
100 kHz	22	32	44
1 MHz	32	44	—
10 MHz	44	—	—

1.6 REACTANCE AND IMPEDANCE FORMULAS

Reactance (X):

1. The concept of reactance applies to sinusoidal waveforms only. Other waveforms must be treated as a fundamental plus harmonic sine waves. Reactive networks will affect each of these components differently, generally resulting in distortion of the waveshape.

2. Reactive devices store energy on the first quarter-cycle of ac and reflect it back to the source on the second quarter cycle. They limit current but do not dissipate power.

3. The reactance of an inductor varies directly with frequency. The reactance of a capacitor varies inversely with frequency.

4. The current and voltage in a purely reactive device are 90° out of phase: v leads i in an inductor, i leads v in a capacitor.

Reactance of inductors (X_L) and of capacitors (X_C) is obtained by:

$$X_L = 2\pi f L \qquad X_C = \frac{-1}{2\pi f C}$$

18

Ohm's law for reactance:

$$I = \frac{V}{X} \qquad X = \frac{V}{I} \qquad V = IX$$

Susceptance is negative reciprocal reactance:

$$B = \frac{-1}{X}$$

Reactances in series add algebraically. Note that capacitive reactances subtract from inductive reactances.

$$X_T = X_1 + X_2 + X_3 + \cdots$$

Reactances in parallel can be calculated by the "reciprocal of the reciprocals" formula. This amounts to changing the reactances to susceptances, adding, and changing the sum back to a reactance.

$$X_T = \frac{1}{\dfrac{1}{X_1} + \dfrac{1}{X_2} + \dfrac{1}{X_3} + \cdots}$$

Resonance (tuned circuit) frequency: Series L-C combination has zero reactance at resonant frequency f_r. Parallel L-C combination has infinite reactance at f_r.

$$f_r = \frac{1}{2\pi \sqrt{LC}}$$

A **reactance chart** (Figure 1.12 on the next two pages) can be used to estimate resonant frequency, and to find reactance at any frequency for any value of capacitor or inductor. The example point shows that 1 µF and 0.5 H resonate at about 230 Hz, and that these components each have a reactance of about 700 Ω at 230 Hz.

Impedance (Z) is the combination of resistance and reactance. The phase angle θ between voltage and current is between 0 and ±90°, and the power dissipated is between IV and 0.

19

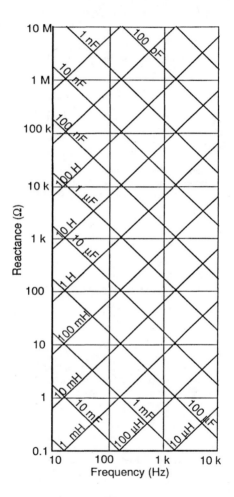

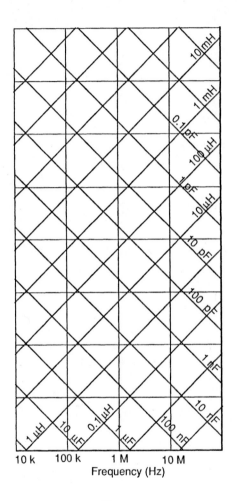

Frequency (Hz)

Power factor (F_P) is the ratio of true power P (given by $I^2 R$) dissipated in the resistance to the "apparent power" S, which is the $I\,V$ product for the circuit, neglecting the phase difference between I and V. In general:

$$F_P = \cos\theta = \frac{P}{S}$$

where θ is the phase angle between circuit voltage and current.

Admittance (Y) is reciprocal impedance:

$$Y = \frac{1}{Z}$$

Ohm's law for impedance:

$$I = \frac{V}{Z} \qquad Z = \frac{V}{I} \qquad V = IZ$$

$$P = IV\cos\theta \qquad P = I^2 Z \cos\theta \qquad P = \frac{V^2}{Z}\cos\theta$$

Phasor (vector) representation of series $R\cdot X$ circuits is shown in Figure 1.13, below.

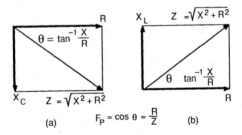

(a)

$$F_P = \cos\theta = \frac{R}{Z} \qquad \text{(b)}$$

Figure 1.13 Phasor representation for series R-C circuits (a), and series R-L circuits (b).

Phasor (vector) representation of parallel *R-X* circuits appears in Figure 1.14, below.

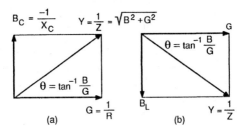

Figure 1.14 Phasor representation of parallel *R-C* (a) and parallel *R-L* circuits (b).

Series-parallel *R-X* conversion. Any series *R-X* circuit can be converted to an equivalent parallel *R-X* circuit, and vice versa. The conversion is valid only at the frequency for which *X* was determined. An example of each conversion is given in Figure 1.15, below.

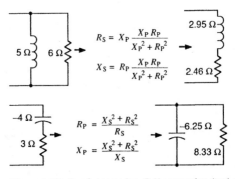

Figure 1.15 Parallel-to-series *R-X* conversion (top) and series-to-parallel *R-X* conversion (below).

R-L-C networks can be used to transform a resistance R up or down (at a single frequency) without the use of a transformer. (See Figure 1.16.) Note that the frequency at which the reactive components cancel is not given by $1/(2\pi\sqrt{LC})$, unless R is very much smaller than X.

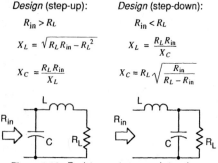

Design (step-up):	*Design (step-down):*

$$R_{in} > R_L \qquad\qquad R_{in} < R_L$$

$$X_L = \sqrt{R_L R_{in} - R_L^2} \qquad X_L = \frac{R_L R_{in}}{X_C}$$

$$X_C = \frac{R_L R_{in}}{X_L} \qquad X_C = R_L\sqrt{\frac{R_{in}}{R_L - R_{in}}}$$

Figure 1.16 Resistance step-up and step-down.

Analysis (both circuits):

$$R_{in} = X_C \sqrt{\frac{R_L^2 + X_L^2}{R_L^2 + (X_L - X_C)^2}}$$

$$R_{in} \approx \frac{X^2}{R_L} = QX \quad \text{[approximate*]}$$

*Q at resonance; error > 2% for $Q > 5$

The Pi network, shown in Figure 1.17, provides a somewhat wider range of impedance transformations without the use of transformers, and is popular in radio transmitters. In the equations that follow, K is a factor used only to facilitate calculation of the other parameters. Load resistance R_L is transformed to input resistance R_{in}. Circuit Q is generally chosen in the range from 8 to 20; higher values tend to eliminate harmonics, and lower values tend to make the circuit broad-banded.

$$K = \frac{R_{in}Q - \sqrt{R_{in}R_L Q^2 - (R_{in} - R_L)^2}}{R_{in} - R_L}$$

$$X_{C1} = \frac{R_{in}}{Q} \qquad\qquad X_{C2} = \frac{R_L}{Q - K}$$

$$X_L = \frac{R_{in}Q}{K^2 + 1}$$

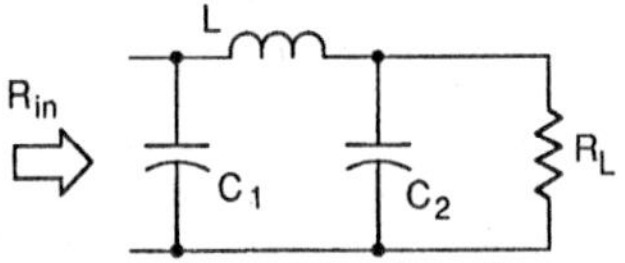

Figure 1.17 Pi network for impedance matching

For example, to transform a 50-Ω resistive load up to 2000 Ω with a pi network having a Q of 12 at 7 MHz:

$$K = \frac{2000 \times 12 - \sqrt{2000 \times 50 \times 12^2 - (2000 - 50)^2}}{2000 - 50} = 10.64$$

$$X_{C1} = \frac{2000}{12} = 167\ \Omega \quad (136\ \mathrm{pF})$$

$$X_{C2} = \frac{50}{12 - 10.64} = 36.8\ \Omega \quad (618\ \mathrm{pF})$$

$$X_L = \frac{2000 \times 12}{10.64^2 + 1} = 210\ \Omega \quad (4.78\ \mu\mathrm{H})$$

Bandwidth, Q, and D: Q is the ratio of energy stored to energy dissipated per cycle in a device or circuit. Dissipation factor D is its reciprocal.

For series R-X elements: $\qquad Q = \dfrac{X_S}{R_S}$

For parallel R-X elements: $\qquad Q = \dfrac{R_P}{X_P}$

For a series-resonant circuit:

$$Q = \frac{X_S}{R_S} = \frac{X_L}{R_S}$$

For a parallel-resonant circuit with parallel resistance:

$$Q = \frac{R_P}{X_C} = \frac{X_P}{X_L}$$

For a parallel-resonant circuit with series resistance in the inductor an approximation to within 4% for $Q_{coil} > 5$ is

$$Q = \frac{X}{R_S}$$

The Q of a tuned circuit provides an estimate of its *bandwidth (B)*, which is the frequency span between the upper and lower half-power points, as shown in Figure 1.18, below, for bandpass and trap circuits.

$$B = \frac{f_r}{Q}$$

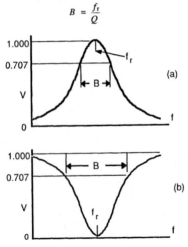

Figure 1.18 Bandpass (a) and trap (b) bandwidth.

1.7 COMPLEX NOTATION: RECTANGULAR AND POLAR (PHASOR) FORMS

Rectangular form is $R \pm jX$, where R is resistance in ohms, X is reactance, and $j = \sqrt{-1}$. R is called the *real* part and X the *imaginary* part. Three circuits and their complex expressions are shown in Figure 1.19 as examples.

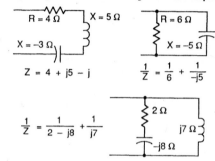

Figure 1.19 Complex expressions for three circuits.

Add and subtract real and imaginary parts separately:

$$(3 + j6) + (2 - j4) = 5 + j2$$

Multiply and divide both parts by real numbers as usual:

$$3(1 + j2) = 3 + j6 \qquad \frac{4 - j6}{-2} = -2 + j3$$

Multiply or divide by a j **term:** remember that $j^2 = -1$, $j^3 = -j$, and $j^4 = +1$.

$$-j3(5 + j) = -j15 - j^2 3 = 3 - j15$$

$$\frac{9 - j3}{j3} = \frac{j^4 9 - j3}{j3} = j^3 3 - 1 = -1 - j3$$

Multiply complex terms by conventional algebra: Product of *first* terms + product of *outer* terms + product of *inner* terms + product of *last* terms (the *foil* method):

$$(8 + j3)(2 - j5) = 16 - j40 + j6 - j^2 15 = 31 - j34$$

27

Divide complex terms by multiplying both terms by the *complex conjugate* of the denominator. This converts the denominator to a real number. Form the complex conjugate by changing the sign of the j term.

$$\frac{5+j10}{1-j2} \times \frac{1+j2}{1+j2} = \frac{5+j10+j10+j^2 20}{1-j^2 4}$$

$$= \frac{-15+j20}{5} = -3+j4$$

Conversion to polar form makes the operations of multiplication and division easier than they are in complex (rectangular) form. Polar form expresses magnitude and phase angle, whereas rectangular form expresses real (resistive or in-phase) and imaginary (reactive or 90°-out-of-phase) components. The equivalency of rectangular and polar forms is expressed as $R + jX = Z \angle\theta$.

To convert rectangular to polar form:

$$Z = \sqrt{R^2 + X^2} \qquad \theta = \tan^{-1}\frac{X}{R}$$

To convert polar to rectangular form:

$$R = Z\cos\theta \qquad\qquad X = Z\sin\theta$$

To multiply polar quantities, multiply the magnitudes but *add* the phase angles algebraically:

$$2 \angle30° \times 4 \angle{-25°} = 8 \angle5°$$

To divide polar quantities, divide the magnitudes but subtract the denominator phase angle from the numerator phase angle algebraically.

$$\frac{12 \angle15°}{4 \angle{-30°}} = 3 \angle45°$$

Using the example from the previous complex division:

$$\frac{5+j10}{1-j2} = \frac{11.18 \angle63.4°}{2.24 \angle{-63.4°}} = 5 \angle126.9° = -3+j4$$

1.8 THREE-PHASE FORMULAS

Power in a 3-phase balanced load is calculated as

$$P_t = 3P_a = \sqrt{3}\, I_L V_L \cos \theta_L$$

where P_t is total power, P_a is the power in one of the three arms of the load, I_L and V_L are current and voltage for one phase (either Δ or Y connected), and θ_L is the angle between I_L and V_L.

Power measurement in a 3-phase unbalanced load by the two-wattmeter method is illustrated in Figure 1.20, below.

$$P_t = P_1 + P_2 \quad \text{where}$$

$$P_1 = I_A V_{AB} \cos \theta_{IA\text{-}VAB} \quad \text{and}$$

$$P_2 = I_C V_{CB} \cos \theta_{IC\text{-}VCB}$$

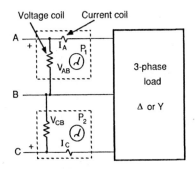

Figure 1.20 Two-wattmeter method of measuring power in a 3-phase system.

Note: If the load is highly unbalanced, one wattmeter may read negative. Reverse its voltage-coil connections and *subtract* its reading from the other meter reading.

1.9 TRANSFORMER FORMULAS

Turns ratio: $n = N_s / N_p$, where N_s and N_p are the number of turns on the secondary and primary, respectively.

Voltage and current ratios:

$$n = \frac{V_s}{V_p} = \frac{I_p}{I_s}$$

Impedance ratio, ideal transformer, unity coupling ($k = 1$):

$$\frac{Z_s}{Z_p} = n^2 \left(\frac{V_s}{V_p}\right)^2$$

where Z_p is the impedance reflected across to the primary when the secondary load is Z_s, and Z_s is the source impedance seen looking back into the secondary when the primary source impedance is Z_p. Figure 1.21, below, illustrates.

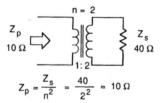

$$Z_p = \frac{Z_s}{n^2} = \frac{40}{2^2} = 10\ \Omega$$

Figure 1.21 (a) A step-up transformer has lower Z_{in}.

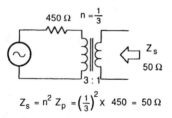

$$Z_s = n^2 Z_p = \left(\frac{1}{3}\right)^2 \times 450 = 50\ \Omega$$

Figure 1.21 (b) Step-down transformer makes Z_{source} appear lower.

Low-frequency cutoff of a transformer:

$$f_{low} = \frac{r_g R_L}{2\pi L_p (n^2 r_g + R_L)}$$

where f_{low} is the –3 dB point, r_g is generator source resistance, R_L is secondary load resistance, L_p is primary inductance in henries, and n is turns ratio, N_s/N_p.

Transformer saturation is observed on an oscilloscope as gross distortion of the secondary waveform from that applied by the source. For a sine-wave driving signal:

$$\frac{V_{1\,(max)}}{f_1} = \frac{V_{2\,(max)}}{f_2}$$

Thus, a transformer that saturates at 15 V at 6 Hz will saturate at 150 V at 60 Hz.

For pulse-driven transformers:

$$V_1 t_1 = V_2 t_2$$

Thus, a pulse transformer that can take 5 V for 30 μs can take 10 V for 15 μs.

1.10 POWER-SUPPLY FORMULAS

Ripple: percent rms, sine-wave-shaped:

$$\eta_{rip} = \frac{V_{o\,(rms\,rip)}}{V_{O\,(dc\,FL)}} \times 100\%$$

$$= \frac{V_{o\,(p\text{-}p\,rip)}}{2.828 V_{O\,(dc\,FL)}} \times 100\%$$

Ripple: percent rms, sawtooth-shaped:

$$\eta_{rip} = \frac{V_{o\,(p\text{-}p\,rip)}}{3.47 V_{O\,(dc\,FL)}} \times 100\%$$

Load regulation: percent, full-load (FL) to no-load (NL):

$$\eta_{reg} = \frac{V_{O\,(NL)} - V_{O\,(FL)}}{V_{O\,(FL)}} \times 100\%$$

Line regulation, $V_{line(max)}$ to $V_{line(min)}$:

$$\eta_{line\ reg} = \frac{Percent\ load\ voltage\ change}{Percent\ line\ voltage\ change} \times 100\%$$

$$= \frac{V_{line\ min}\ (V_{LOAD\ max} - V_{LOAD\ min})}{V_{LOAD\ min}\ (V_{line\ max} - V_{line\ min})} \times 100\%$$

1.11 DECIBELS

Decibels express a power ratio, not an amount. They tell how many times more (positive dB) or less (negative dB) but not how much in absolute terms. Decibels are logarithmic, not linear. For example, 20 dB is not twice the power ratio of 10 dB. The defining equation for decibels is

$$\alpha_{dB} = 10 \log \frac{P_2}{P_1} \quad [units\ of\ dB]$$

where P_2 is the power being measured, and P_1 is the reference to which P_2 is being compared. To convert from decibel measure back to power ratio:

$$\frac{P_2}{P_1} = \log^{-1}\left(\frac{\alpha_{dB}}{10}\right) = 10^{(\alpha_{dB}/10)}$$

Voltage is more easily measured than power, making it generally more convenient to use:

$$\alpha_{dB} = 20 \log \frac{V_2}{V_1} \quad [Z_2 = Z_1]$$

The equation for obtaining voltage ratio from dB is

$$\frac{V_2}{V_1} = \log^{-1}\left(\frac{\alpha_{dB}}{20}\right) = 10^{(\alpha_{dB}/20)} \quad [Z_2 = Z_1]$$

Decibels are defined in terms of power ratios. Note well that the voltage-ratio equations are valid *only* if the two voltages appear across equal impedances. However, in audio systems where Z_o is essentially zero and Z_{in} is essentially infinite, it is common to use the voltage equations without regard to impedances. If this is done, the decibel values obtained must in no case be applied to power or power-gain calculations.

Decibel to Voltage-ratio to Power-ratio Table

α_{dB}	$\dfrac{V_2}{V_1}$	$\dfrac{P_2}{P_1}$	α_{dB}	$\dfrac{V_2}{V_1}$	$\dfrac{P_2}{P_1}$
0	1.000	1.000	30	31.62	1000
0.5	1.059	1.122	32	39.81	1585
1	1.122	1.259	34	50.12	2512
2	1.259	1.585	36	63.10	3981
3	1.413	1.995	38	79.43	6310
4	1.585	2.512	40	100.0	10 000
5	1.778	3.162	42	125.9	15 850
6	1.995	3.981	44	158.5	25 120
7	2.239	5.012	46	199.5	39 810
8	2.512	6.310	48	251.2	63 100
9	2.818	7.943	50	316.2	1.00×10^5
10	3.162	10.00	52	398.1	1.58×10^5
11	3.548	12.59	54	501.2	2.51×10^5
12	3.981	15.85	56	631.0	3.98×10^5
13	4.467	19.95	58	794.3	6.31×10^5
14	5.012	25.12	60	1000	1.00×10^6
15	5.623	31.62	62	1259	1.58×10^6
16	6.310	39.81	64	1585	2.51×10^6
17	7.079	50.12	66	1995	3.98×10^6
18	7.943	63.10	68	2512	6.31×10^6
19	8.913	79.43	70	3162	1.00×10^7
20	10.00	100.0	75	5623	3.16×10^7
22	12.59	158.5	80	10 000	1.00×10^8
24	15.85	251.2	85	17 780	3.16×10^8
26	19.95	398.1	90	31 620	1.00×10^9
28	25.12	631.0	95	56 230	3.16×10^9

Negative decibels represent loss factors (division). Thus a 6-dB amplifier has a voltage gain of 2, but an attenuator that cuts voltage down to $\frac{1}{2}$ is a –6-dB attenuator.

Zero-dB standards:

Audio industry: 0 dB = 1 mW in 600-Ω resistance
                                      = 0.7746 V rms across 600 Ω

Measurements to this standard use the 'unit symbol' dBm.

Television industry: 0 dB = 1 mV rms across 75 Ω
                                        = 1.333×10^{-8} W

As an example, a microphone rated at −52 dB delivers a power P_2 of

$$P_2 = P_1 \; \log^{-1}\left(\frac{-52}{10}\right) = 1 \text{ mW} \times 6.3 \times 10^{-6} = 6.3 \text{ nW}$$

Adding decibels is equivalent to multiplying voltage ratios. Subtracting decibels is equivalent to dividing voltage ratios.

    Voltage ratios: $A_v = 2 \times 10 \div 4 = 5$ times

    Decibels $\alpha_{dB} = 6 + 20 - 12 = 14$ dB

1.12 INSTRUMENTATION DEFINITIONS

Precision. The degree to which a measurement is readable or is specified. May be indicated in units of measure (example: *to within ±10 mV*) or in percent (example: *readable to within 0.5% of full scale*).

Accuracy. The degree to which the indicated value approaches the true value. Usually expressed in terms of *error*, below.

Error. The difference between true and indicated values of the measured quantity. Often expressed as a percent of full range of the instrument.

$$\varepsilon = \frac{V_{\text{indicated}} - V_{\text{true}}}{V_{\text{true}}} \times 100\%$$

In instrumentation work the word *error* should be reserved for unavoidable inaccuracies in the measurement process. The word *mistake* should be used for avoidable blunders.

34

Resolution. The smallest increment that will render one reading distinguishable from another.

Sensitivity. The ratio of output response to input stimulation. Often expressed as input required for full-scale (f.s.) output, or input required for minimum observable output.

Linearity. The degree to which the graph of input stimulation vs. output response approaches a straight line. Expressed by percent *nonlinearity*, as illustrated in Figure 1.22, below. *Normal linearity* has the end points of the ideal line and the actual curve coincident, as shown below. *Zero-based* linearity has the zero points coincident while the slope of the ideal line is adjusted for lowest percent deviation from the actual curve. *Independent* or *floating* linearity allows both end points to be adjusted for lowest percent nonlinearity. The deviations are commonly expressed as a percent of full scale.

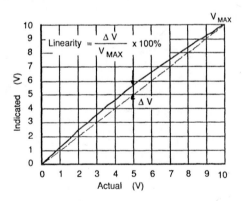

Figure 1.22 Illustration of determination of normal or end-point linearity.

1.13 ATTENUATOR FORMULAS

Pads (Figures 1.23 and 1.24) are attenuators that preserve the line impedance looking at them from either direction.

Definitions: $\quad A_v = \dfrac{V_{out}}{V_{in}} \qquad a = \dfrac{V_{in}}{V_{out}} = \dfrac{1}{A_v}$

$$Z_o = R_s = R_L \qquad \text{(line, source , and load)}$$

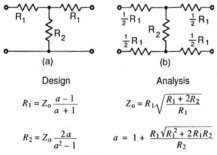

(a)

Design

$$R_1 = Z_o \frac{a^2 - 1}{2a}$$

$$R_2 = Z_o \frac{a + 1}{a - 1}$$

(b)

Analysis

$$Z_o = \frac{R_2}{\sqrt{\dfrac{R_1 + 2R_2}{R_1}}}$$

$$a = \frac{R_1 + R_2 \parallel R_1}{R_2 \parallel R_1}$$

Figure 1.23 (a) Pi pad, and (b) O-pad.

(a)

Design

$$R_1 = Z_o \frac{a - 1}{a + 1}$$

$$R_2 = Z_o \frac{2a}{a^2 - 1}$$

(b)

Analysis

$$Z_o = R_1 \sqrt{\frac{R_1 + 2R_2}{R_1}}$$

$$a = 1 + \frac{R_1 \sqrt{R_1^2 + 2R_1 R_2}}{R_2}$$

Figure 1.24 (a) Tee-pad, and (b) H-pad.

Padding a line reduces impedance mismatch when the line is open-circuited or short-circuited. The graph of Figure 1.25 shows how higher attenuations cause an open line to look more like the characteristic impedance Z_o at the pad input R_{in}. The ratio Z_o/R_{in} is similar for a shorted line.

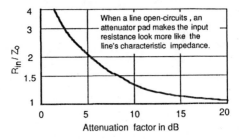

When a line open-circuits, an attenuator pad makes the input resistance look like the line's characteristic impedance.

Figure 1.25 Effect of padding a line.

Frequency-compensated attenuators (Figure 1.26) maintain the same attenuation ratio at high frequencies, where the reactance of stray capacitance would ordinarily upset the resistance ratio.

$$a = \frac{V_{in}}{V_o} = \frac{R_1 + R_2}{R_1}$$

$$R_1 C_1 = R_2 C_2$$

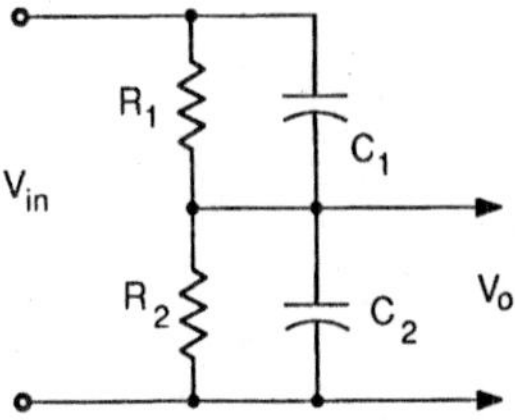

Figure 1.26 Frequency-compensated attenuator.

An **L-pad** (Figure 1.27) can be used to control the output to a load while maintaining a constant input impedance. Note that R_1 is a linear potentiometer, and R_2 is logarithmic.

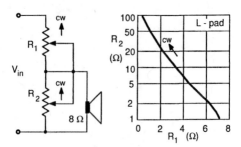

Figure 1.27 L-pad and required resistances.

1.14 BRIDGE CIRCUITS

The **potentiometer circuit** (Figure 1.28) is used to measure unknown voltages V_x without loss across a high-value internal resistance R_x.

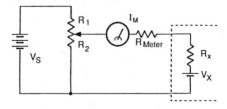

Figure 1.28 Potentiometer. Adjust for $I = 0$.

$$V_X = V_S \frac{R_2}{R_1 + R_2} \qquad I_M = \frac{\dfrac{V_S R_2}{R_1 + R_2} - V_X}{R_1 \| R_2 + R_M + R_X}$$

The **Wheatstone bridge** (Figure 1.29) is used to detect small changes in one of the four resistances. At balance:

$$\frac{R_1}{R_2} = \frac{R_3}{R_4}$$

If all resistances are nominally equal, and one resistor changes by ΔR:

$$I_M = \frac{V_S \Delta R}{4R(R + R_M)}$$

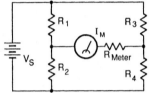

R_1 R_3

I_M

R_{Meter}

V_S

R_2

R_4

Figure 1.29

Wheatstone Bridge:

Adjust for $I_M = 0$.

For a slightly unbalanced Wheatstone bridge where the resistance arms are not equal, and arm R_4 changes by ΔR:

$$I_M = \frac{V_S R_3 \Delta R}{(2R_3R_4 + R_3^2 + R_4^2)(R_1 \| R_2 + R_3 \| R_4 + R_M)}$$

The **Kelvin bridge** (Figure 1.30) is used to measure very low value resistances. The balance equations are the same as for the Wheatstone bridge. The effects of the flexible lead wires to unknown resistance R_x are eliminated by R_5 and R_6, which should be chosen so that

$$\frac{R_5}{R_6} = \frac{R_1}{R_2}$$

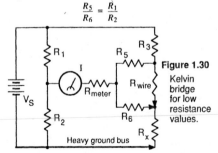

R_1 R_5 R_3

I

R_{meter} R_{wire}

V_S

R_2 R_6

R_x

Heavy ground bus

Figure 1.30

Kelvin bridge for low resistance values.

The **series-capacitance comparison bridge** (Figure 1.31) can be used to measure an unknown capacitor C_S with a low dissipation factor D. R_S represents losses within the capacitor. The 'meter' is any sensitive ac detector. R_1 may be calibrated in values of C_S, and R_2 in values of D.

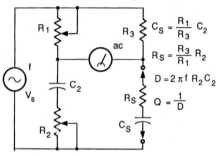

Figure 1.31   Series-capacitance comparison bridge.

$$C_S = \frac{R_1}{R_3} C_2$$

$$R_S = \frac{R_3}{R_1} R_2$$

$$D = 2\pi f R_2 C_2$$

$$Q = \frac{1}{D}$$

The **Hay bridge** (Figure 1.32) is used to determine a parallel-equivalent circuit for high-Q inductors. To convert to a series-equivalent circuit use:

$$L_s = \frac{R_2 R_3 C_1}{1 + 2\pi f R_1^2 C_1^2} \qquad R_s = \frac{2\pi f R_1 R_2 R_3 C_1}{1 + 2\pi f R_1^2 C_1^2}$$

$$L_P = R_2\, R_3\, C_1$$

$$R_P = \frac{R_3}{R_1} R_2$$

$$D = 2\pi f R_1 C_1$$

$$Q = \frac{1}{D}$$

Figure 1.32   Hay bridge for high-Q inductors.

The Maxwell bridge (Figure 1.33) is used to measure inductances with lower-Q, in the vicinity of 0.1 to 5.

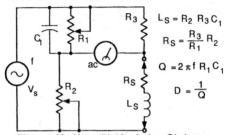

$$L_S = R_2\, R_3\, C_1$$

$$R_S = \frac{R_3}{R_1}\, R_2$$

$$Q = 2\pi f\, R_1\, C_1$$

$$D = \frac{1}{Q}$$

Figure 1.33 Maxwell bridge for low-Q inductors.

The Wien bridge (Figure 1.34) is a notch filter with high attenuation at f_c and 6-dB attenuation at $0.5 f_c$ and $2 f_c$ (if the resistance at V_o is very high.) Both sides of the output must remain isolated from the cold side of the source.

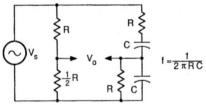

$$f = \frac{1}{2\pi R C}$$

Figure 1.34 Wien bridge notch filter.

The twin-tee notch filter (Figure 1.35) is similar to the Wien bridge, but one side of the output is grounded.

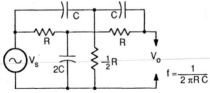

$$f = \frac{1}{2\pi R C}$$

Figure 1.35 Twin-tee notch filter.

41

1.15 FILTERS

***R-C* low-pass filters** are shown in Figure 1.36, below.
Output is down 3 dB (× 0.707) at f_c, 20 dB (× 0.1) at 10 f_c,
and 40 dB (× 0.01) at 100 f_c.

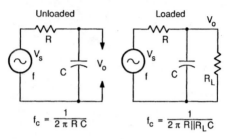

Figure 1.36 *R-C* low-pass filters.

An ***R-C* high-pass filter** is shown in Figure 1.37, below,
with a sketch of the Bode plot of its output vs. frequency. Its
output is –3 dB at f_c, –20 dB at 0.1 f_c, and –40 dB at 0.01
f_c, similar to the *R-C* low-pass filter.

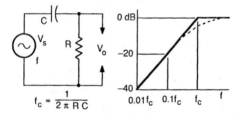

Figure 1.37 High-pass filter and frequency-response.

Cascaded *R-C* filter sections can be employed to increase
attenuation and roll-off rate. Attenuation for two sections is
–6 dB at f_c , –12 dB at 2 f_c (or 1/ 2 f_c), and 40 dB at 10
f_c (or 0.1 f_c).

42

A bandpass R-C filter can be made by cascading a high-pass section after a low-pass section. Let the resistors in both sections be equal, and keep f_{high} at least 10 x f_{low}.

L-C high- and low-pass filters have roll-offs from f_c that are three times as steep as RC filters. Attenuation for the single-section circuits shown in Figures 1.38 and 1.39 is -3 dB at f_c, and -18 dB for each factor of two away from f_c. Sections may be cascaded for steeper roll-off and greater attenuation outside the passband. The relevant equations are

Analysis	Design
$Z_0 = \sqrt{\dfrac{L}{C}}$	$L = \dfrac{Z_0}{2\pi f_c}$
$f_c = \dfrac{1}{2\pi\sqrt{LC}}$	$C = \dfrac{1}{2\pi f_c Z_0}$

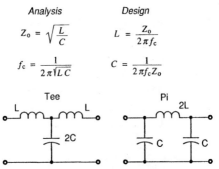

Figure 1.38 Constant-k low-pass filters.

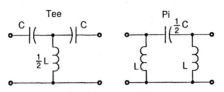

Figure 1.39 Constant-k high-pass filters.

End sections added to L-C filters preserve a more constant impedance over the passband and place a notch just into the reject band from f_c, resulting in a very steep roll-off. Two end sections are shown in Figures 1.40 and 1.41, on page 44.

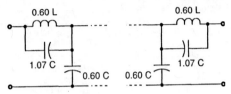

Figure 1.40 End sections for Pi low-pass filters.

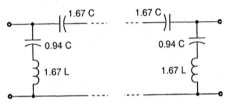

Figure 1.41 End sections for Tee high-pass filters.

Bandpass L-C filters pass a wider range of frequencies than a simple tuned circuit. The generator resistance r_g and the load resistance R_L must each equal the characteristic impedance Z_0 of the filter. The ratio of high to low –3-dB frequencies (f_1/f_2) should generally be in the range of 1.2 to 5.0 to avoid severe attenuation in the passband. The same equations apply to Figures 1.42 and 1.43.

$$L_1 = \frac{Z_0}{\pi(f_2 - f_1)} \qquad C_1 = \frac{f_2 - f_1}{4\pi Z_0 f_1 f_2}$$

$$L_2 = \frac{Z_0(f_2 - f_1)}{4\pi f_1 f_2} \qquad C_2 = \frac{1}{\pi Z_0(f_2 - f_1)}$$

Figure 1.42 Tee bandpass filter.

44

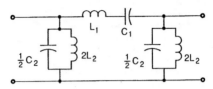

Figure 1.43 Pi bandpass filter.

Bandstop L-C filters eliminate a relatively wide range of frequencies while passing those below f_1 or above f_2. The same equations apply to Figures 1.44 and 1.45.

$$L_1 = \frac{Z_0(f_2 - f_1)}{\pi f_1 f_2} \qquad C_1 = \frac{1}{4\pi Z_0 (f_1 - f_2)}$$

$$L_2 = \frac{Z_0}{4\pi(f_2 - f_1)} \qquad C_2 = \frac{f_2 - f_1}{\pi Z_0 (f_1 f_2)}$$

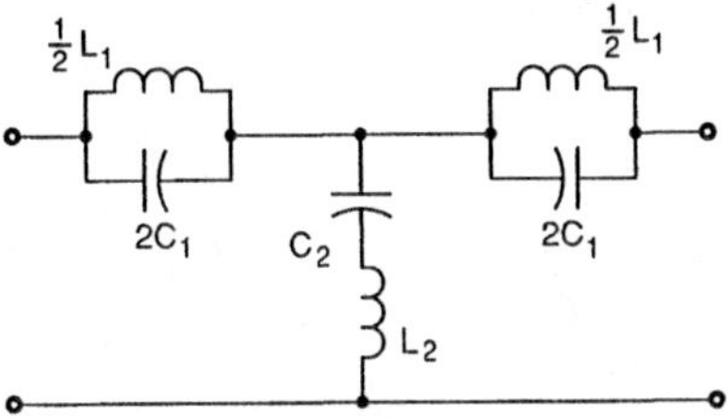

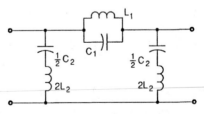

Figure 1.44 Tee bandstop filter.

Figure 1.45 Pi bandstop filter.

Active filters use R-C elements and op amps to achieve controlled roll-off characteristics as well as voltage gain. Figure 1-46 shows three simple active filters for audio frequencies. Higher values of R_v in (a) and (b) will produce roll-offs sharper than 40 dB/decade, but will introduce pulse distortion. In (c), varying R3 will vary the peak frequency f_c.

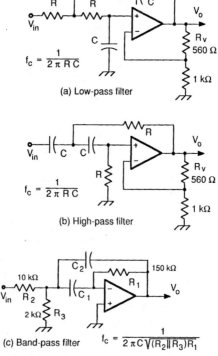

$$f_c = \frac{1}{2\pi RC}$$

(a) Low-pass filter

$$f_c = \frac{1}{2\pi RC}$$

(b) High-pass filter

$$f_c = \frac{1}{2\pi C\sqrt{(R_2\|R_3)R_1}}$$

(c) Band-pass filter

Figure 1.46 Active filters using op amps.

1.16 ANTENNAS AND TRANSMISSION LINES

Frequency and Wavelength. The physical length of a line is less than its electrical length by the *velocity factor*, which is approximately 0.975 for open wires, 0.83 for 300-Ω twin lead, and 0.7 to 0.8 for coax cables. Electrical wavelength is

$$\lambda_{meters} = \frac{300 \times 10^6}{f_{Hz}} \quad \text{or} \quad \lambda_{feet} = \frac{984 \times 10^6}{f_{Hz}}$$

Voltage received by a dipole antenna, where V_r is received voltage, P is transmitted power, G_1 and G_2 are the gains of the transmitting and receiving antennas (about 1.6 each for dipoles broadside), λ is the wavelength in meters, d is the distance between antennas in meters, and Z is antenna impedance. Of course, the formula assumes a free-space environment with no obstructions:

$$V_r = \sqrt{\frac{P G_1 G_2 \lambda^2 Z}{16 \pi^2 d^2}}$$

Characteristic impedance of a line:

$$Z_0 = \sqrt{\frac{L}{C}} \quad \text{where } L \text{ and } C \text{ are per unit length.}$$

For air-insulated twin lead: $\quad Z_0 = 276 \log \frac{w}{t}$

where w is center-to-center width between wires and t is thickness (diameter) of one wire.

For coaxial cables: $\quad Z_0 = 138 \log \frac{d}{t \sqrt{K}}$

where d is inside diameter of the outer conductor and K is the dielectric constant of the insulation.

Quarter-wave impedance transformer (resistive load):

$$Z_{in} = \frac{Z_0^2}{R_L} \quad \text{or} \quad Z_0 = \sqrt{Z_{in} R_L}$$

Reflection coefficient and standing wave ratio for lines with load resistance R_L shunted by reactance X_L:

$$\Gamma = \sqrt{\frac{(R_L - Z_0)^2 + X_L^2}{(R_L + Z_0)^2 + X_L^2}} \quad \text{and} \quad SWR = \frac{1 + \Gamma}{1 - \Gamma}$$

For purely resistive loads, $SWR = \dfrac{R_L}{Z_0}$ or $SWR = \dfrac{Z_0}{R_L}$

Input impedance for line of electrical length l (meters), load $X_L \| R_L$, at frequency f (in MHz):

$$R_{in} = \frac{R_L(1+k^2)}{\left(1 - k\frac{X_L}{Z_0}\right)^2 + \left(k\frac{R_L}{Z_0}\right)^2} \qquad X_{in} = \frac{X_L(1-k^2) + k\left(1 - \frac{R_L^2 - X_L^2}{Z_0^2}\right)}{\left(1 - k\frac{X_L}{Z_0}\right)^2 + \left(k\frac{R_L}{Z_0}\right)^2}$$

where $k = \tan 1.2 f l$

Doppler frequency shift for electromagnetic waves, where f_o is observed frequency, f_s is source frequency, v is relative velocity of source and observer, and c is the speed of light (300 Mm/s):

$$f_o = f_s \sqrt{\frac{c+v}{c-v}}$$

Doppler frequency shift for sound waves, where f_o is observed frequency, f_s is source frequency, v is velocity of sound, v_o is velocity of observer, and v_s is velocity of source, all in meters per second:

$$f_o = f_s \frac{v + v_o}{v - v_s}$$

1.17 THERMAL FORMULAS

Ohm's Law Analogy: Thermal power (P, in watts) flows through thermal resistance R_θ, producing a temperature differential ΔT. Significant thermal resistances are $R_{\theta(J\text{-}C)}$ (junction-to-case), $R_{\theta(C\text{-}S)}$ (case-to-sink), and $R_{\theta(S\text{-}A)}$ (sink-to-ambient).

$$P = \frac{\Delta T}{R_\theta} = \frac{T_{J(max)} - T_A}{R_{\theta(J\text{-}C)} + R_{\theta(C\text{-}S)} + R_{\theta(S\text{-}A)}}$$

The thermal resistances of common objects in still, open air are given in the tables on the following page.

Thermal resistances: black metal plate

Area of one side (cm^2)	$R_{\theta(S-A)}$
12	30
20	14
30	8
60	5
150	3

Thermal resistances: unmounted transistor cases

TO-92 plastic flat front	300
TO-18 mini TO-5 metal	300
TO-5 standard round metal	150
TO-60 stud mount	70
TO-66 mini TO-3 (dime-sized)	60
TO-220 power tab mount	50
TO-3 standard power (quarter-sized)	30
TO-36 1^1/4-inch round	25

Temperature rise vs. resistance. The temperature rise of transformers, motors, or any device containing metallic resistances can be calculated by the resistance change from cold to hot:

$$\frac{R_H}{R_C} = K^{(T_H - T_C)} \qquad T_H - T_C = \frac{\log(R_H/R_C)}{\log K}$$

where R is resistance, T is temperature in °C, K is the temperature coefficient of resistance of the metal per °C (see table below) and H and C refer to *hot* and *cold*, respectively.

Temperature coefficient K (per °C) for various metals

Aluminum	1.0040
Copper	1.0039
Gold	1.0034
Iron	1.0055
Nichrome	1.00017
Silver	1.0038
Tungsten	1.005

Resistance ratios vs. temperature differentials (°C) for copper

R_H/R_C	$T_H - T_C$	R_H/R_C	$T_H - T_C$
1.04	10	1.37	80
1.08	20	1.48	100
1.12	30	1.60	120
1.17	40	1.72	140
1.21	50	1.86	160
1.26	60	2.02	180
1.31	70	2.18	200

1.18 MATHEMATICAL FORMULAS

Geometric Formulas

Definitions: A = area a = altitude
b = base c = hypotenuse C = circumference
d = diameter h = height r = radius
V = volume s = side of a regular polygon

Circle: $C = 2\pi r = \pi d$ $A = \pi r^2$

Sector: $A = \dfrac{fr}{2} = \pi r^2 \dfrac{\theta°}{360°}$

Segment: $A = \pi r^2 \dfrac{\phi°}{360°} - \dfrac{b(r-h)}{2}$ $b = 2\sqrt{2hr - h^2}$

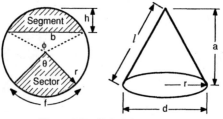

Figure 1.47 Circle (left) and cone.

Sphere: $A = 4\pi r^2 = \pi d^2$ $V = \dfrac{4\pi r^3}{3}$

50

Ellipse: $A = \dfrac{\pi d_1 d_2}{4}$ $[d_1\ \&\ d_2 = \text{major \& minor axes}]$

Cone: $A = \pi r\, l = \pi r \sqrt{r^2 + a^2}$ (base disk excluded)

$$V = \frac{\pi r^2 a}{3}$$

Torus (doughnut with circular cross section; Figure 1.48):

$$A = \pi^2 d D = 4\pi^2 r R$$
$$V = 2.463\, d^2 D = 2\pi^2 r^2 R$$

Figure 1.48 Torus.

Regular pentagon: $A = 1.720\, s^2$

Regular hexagon: $A = 2.598\, s^2$

Regular octagon: $A = 4.828\, s^2$

Regular polygon of n sides: $A = \dfrac{s^2}{4 \tan \dfrac{180°}{n}}$

Trapezoid; Figure 1.49: $A = \dfrac{h\,(a+b)}{2}$

Quadrilateral; Figure 1.49: $A = \dfrac{x\,y \sin\theta}{2}$

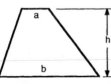

Figure 1.49 Trapezoid (left) and quadrilateral.

Triangle, right: Figure 1-50:

$$A = \tfrac{1}{2}\,b\,a \qquad\qquad c^2 = a^2 + b^2$$

$$\sin A = \frac{a}{c} \qquad\qquad \csc A = \frac{c}{a}$$

$$\cos A = \frac{b}{c} \qquad\qquad \sec A = \frac{c}{b}$$

$$\tan A = \frac{a}{b} \qquad\qquad \cot A = \frac{b}{a}$$

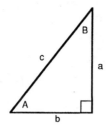 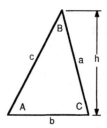

Figure 1.50 Right triangle (left) and oblique triangle.

Triangle, oblique; Figure 1-50: $A = \tfrac{1}{2}\,b\,h$

$$A = \sqrt{m(m-a)(m-b)(m-c)}$$

$$\text{where } m = \frac{a+b+c}{2}$$

$$a = b \cos C + c \cos B$$

Law of sines: $\dfrac{a}{\sin A} = \dfrac{b}{\sin B} = \dfrac{c}{\sin C}$

Law of cosines: $a^2 = b^2 + c^2 - 2bc\,\cos A$

Trigonometric identities:

$$\sin^2 A + \cos^2 A = 1 \qquad\qquad \sec^2 A - \tan^2 A = 1$$

$$\csc^2 A - \cot^2 A = 1$$

$$\sin(-A) = -\sin A \qquad\qquad \cos(-A) = \cos A$$

$$\tan(-A) = -\tan A$$

$$\sin (A + B) = \sin A \cos B + \cos A \sin B$$

$$\cos (A + B) = \cos A \cos B - \sin A \sin B$$

$$\tan (A + B) = \frac{\tan A + \tan B}{1 - \tan A \tan B}$$

$$\sin \frac{A}{2} = \pm \left(\frac{1 - \cos A}{2} \right)^{\frac{1}{2}}$$

$$\cos \frac{A}{2} = \pm \left(\frac{1 + \cos A}{2} \right)^{\frac{1}{2}}$$

$$\tan \frac{A}{2} = \pm \left(\frac{1 - \cos A}{1 + \cos A} \right)^{\frac{1}{2}}$$

Algebra

Determinants for the solution of n linear equations in n unknowns:

Standard form of the equations in two unknowns:
$$a_1 x + b_1 y = k_1$$
$$a_2 x + b_2 y = k_2$$

Solution:

$$x = \frac{\begin{vmatrix} k_1 & b_1 \\ k_2 & b_2 \end{vmatrix}}{\begin{vmatrix} a_1 & b_1 \\ a_2 & b_2 \end{vmatrix}} = \frac{k_1 b_2 - b_1 k_2}{a_1 b_2 - b_1 a_2} \qquad y = \frac{\begin{vmatrix} a_1 & k_1 \\ a_2 & k_2 \end{vmatrix}}{\begin{vmatrix} a_1 & b_1 \\ a_2 & b_2 \end{vmatrix}} = \frac{a_1 k_2 - k_1 a_2}{a_1 b_2 - b_1 a_2}$$

Third-order determinant. Technique not valid for higher orders. Denominator shown. Substitute k values for a values in numerator to solve for x, k values for b to get y, and k values for c to get z.

$$\begin{vmatrix} a_1 & b_1 & c_1 \\ a_2 & b_2 & c_2 \\ a_3 & b_3 & c_3 \end{vmatrix} \begin{matrix} a_1 & b_1 \\ a_2 & b_2 \\ a_3 & b_3 \end{matrix} = \begin{matrix} a_1 b_2 c_3 + b_1 c_2 a_3 + c_1 a_2 b_3 \\ -c_1 b_2 a_3 - a_1 c_2 b_3 - b_1 a_2 c_3 \end{matrix}$$

Fourth-order determinant. $\quad D =$

$$a_1(b_2c_3d_4 + c_2d_3b_4 + d_2b_3c_4 - d_2c_3b_4 - b_2d_3c_4 - c_2b_3d_4)$$
$$- a_2(b_1c_3d_4 + c_1d_3b_4 + d_1b_3c_4 - d_1c_3b_4 - b_1d_3c_4 - c_1b_3d_4)$$
$$+ a_3(b_1c_2d_4 + c_1d_2b_4 + d_1b_2c_4 - d_1c_2b_4 - b_1d_2c_4 - c_1b_2d_4)$$
$$- a_4(b_1c_2d_3 + c_1d_2b_3 + d_1b_2c_3 - d_1c_2b_3 - b_1d_2c_3 - c_1b_2d_3)$$

Determinant simplification. Example in third order:
Multiply column 2 by 3 and subtract from column 1:

$$\begin{vmatrix} 8 & 2 & 5 \\ 9 & 7 & 4 \\ 3 & 1 & 6 \end{vmatrix} = \begin{vmatrix} 8-6 & 2 & 5 \\ 9-21 & 7 & 4 \\ 3-3 & 1 & 6 \end{vmatrix} = \begin{vmatrix} 2 & 2 & 5 \\ -12 & 7 & 4 \\ 0 & 1 & 6 \end{vmatrix}$$

Multiply row 1 by 6 and add to row 2:

$$\begin{vmatrix} 2 & 2 & 5 \\ -12 & 7 & 4 \\ 0 & 1 & 6 \end{vmatrix} = \begin{vmatrix} 2 & 2 & 5 \\ -12+12 & 7+12 & 4+30 \\ 0 & 1 & 6 \end{vmatrix} = \begin{vmatrix} 2 & 2 & 5 \\ 0 & 19 & 34 \\ 0 & 1 & 6 \end{vmatrix}$$

Quadratic formula. Standard form : $ax^2 + bx + c = 0$

$$\text{Solution:} \quad x = \frac{-b \pm \sqrt{b^2 - 4ac}}{2a}$$

Expansions and factors:

$$(a + b)^2 = a^2 + 2ab + b^2 \qquad (a - b)^2 = a^2 - 2ab + b^2$$
$$(a \pm b)^3 = a^3 \pm 3a^2b + 3ab^2 \pm b^3$$
$$(a + b + c)^2 = a^2 + b^2 + c^2 + 2ab + 2ac + 2bc$$
$$(a + b)(a - b) = a^2 - b^2$$
$$(a + jb)(a - jb) = a^2 + b^2$$
$$(a + b)(a^2 - ab + b^2) = a^3 + b^3$$
$$(a - b)(a^2 + ab + b^2) = a^3 - b^3$$

Number of combinations of n things taken r at a time:

$$C_{r\,\text{of}\,n} = \frac{n!}{r!(n-r)!}$$

Number of permutations of n things taken r at a time:

$$P_{r\,\text{of}\,n} = \frac{n!}{(n-r)!}$$

For example, of six items, $a, b, c, d, e,$ and f, taking three at a time, there are 20 combinations but 120 permutations. The two sets a, b, c and a, c, b are different permutations but not different combinations.

The normal curve (gaussian distribution; Figure 1.51):

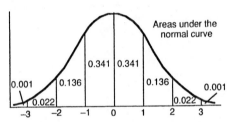

Figure 1.51 Standard deviations from the mean.

Tail areas under the normal curve, one side, vs. standard deviations from the mean

x	A	x	A	x	A
0.0	0.5000	1.0	0.1587	3.0	0.00135
0.1	0.4602	1.2	0.1151	3.1	0.00097
0.2	0.4207	1.4	0.0808	3.2	0.00069
0.3	0.3821	1.6	0.0548	3.3	0.00048
0.4	0.3446	1.8	0.0359	3.4	0.00034
0.5	0.3085	2.0	0.0228	3.5	0.00023
0.6	0.2743	2.2	0.0139	3.6	0.00016
0.7	0.2420	2.4	0.0082	3.8	0.00007
0.8	0.2119	2.6	0.0047	4.0	0.00003
0.9	0.1841	2.8	0.0026	4.2	0.00001

2

Component Data
and Characteristics

2.1 WIRE AND CABLE

Ampacity. Maximum currents (amperes) for copper wire, based on 30°C ambient, 100°C final temperature. A safety margin of 100% recommended: reduce current to half.

AWG Size	Single Wire in Free Air (A)	Bundled Wires, Confined (A)
6	95	55
8	62	39
10	50	31
12	40	23
14	32	17
16	22	13
18	16	10
20	11	7.5
22	7	5
24	3.5	2.1
26	2.2	1.5
28	1.4	0.8
30	0.8	0.5
32	0.5	0.3

Resistance of conductors that are not wire-shaped is determined from

$$R = \frac{\rho l}{A}$$

where ρ is in $\mu\Omega\cdot$cm from the table (next page), l is length in cm, and A is cross-sectional area in cm^2.

Conductivity and Temperature Coefficient of Various Conductors at 20°C

Material	Resistivity Relative to Copper	Resistivity (Ω-cmil/ft)	Resistivity, ρ (μΩ-cm)	Temperature Coefficient per °C at 20°C
Aluminum	1.64	17	2.83	+0.0040
Brass	3.58	37	7.0	+0.0015
Copper	1.00	10.37	1.724	+0.0039
Gold	1.42	14.7	2.44	+0.0034
Iron	5.59	58	9.64	+0.0055
Lead	11.86	123	20.4	+0.0039
Mercury	100	—	172	—
Nichrome	65.09	675	112	+0.0017
Silver	0.945	9.80	1.63	+0.0038
Steel (soft)	9.24	95.8	15.9	+0.0016
Tin	6.68	69.3	11.5	+0.0042
Tungsten	3.20	33.2	5.52	+0.005

Characteristics of Copper Wire at 20°C

Size AWG[1]	Diameter (mils)[2]	Area cmil[3]	Area in.² × 10⁻⁶	Ohms per 1000 ft.[4]	Feet per pound[5]	Turns per in.²[6]	Typical Strandings[7]
0000	460.0	211 600	166 200	0.04901	1.5	—	2104/30
000	409.6	167 800	131 800	1.06182	1.9	—	1661/30
00	364.8	133 100	104 500	0.07793	2.4	—	1330/30
0	324.9	105 600	82 910	0.09825	3.1	—	1045/30
1	289.3	83 690	65 730	0.1239	3.9	—	817/30
2	257.6	66 360	52 120	0.1563	4.9	—	665/30
3	229.4	52 620	41 330	0.1971	6.2	—	—
4	204.3	41 740	32 780	0.2485	7.8	—	133/25
5	181.9	33 090	25 990	0.3134	10	—	—
6	162.0	26 240	20 610	0.3952	12	—	133/27
7	144.3	20 820	16 350	0.4981	16	—	—
8	128.5	16 510	12 970	0.6281	20	—	133/29, 168/30
9	114.4	13 090	10 280	0.7925	25	—	—
10	101.9	10 380	8 155	0.9988	32	88	19/22, 105/30
11	90.7	8 226	6 461	1.260	40	112	—
12	80.8	6 529	5 128	1.590	50	140	19/25, 65/30
13	72.0	5 184	4 072	2.00	63	176	19/25, 65/30
14	64.1	4 109	3 227	2.52	80	220	19/27, 41/30

15	57.1	3 260	2 561	3.18	101	260	—
16	50.8	2 581	2 027	4.02	128	320	19/29, 26/30
17	45.3	2 052	1 612	5.05	160	400	16/30, 65/36
18	40.3	1 624	1 276	6.39	203	500	—
19	35.9	1 289	1 012	8.05	256	630	41/36, 10/30
20	32.0	1 024	804	10.1	323	790	19/33
21	28.5	812	638	12.8	401	990	26/36, 7/30
22	25.3	640	503	16.2	514	1 240	10/32, 16/34
23	22.6	511	401	20.3	644	1 530	16/36, 7/32
24	20.1	404	317	25.7	818	1 890	8/34, 12/36
25	17.9	320	252	32.4	1 031	2 300	10/36, 26/40
26	15.9	253	199	41.0	1 330	2 900	8/36, 7/35
27	14.2	202	158	51.4	1 639	3 700	6/36, 7/36
28	12.6	159	125	65.3	2 067	4 581	5/36
29	11.3	128	100	81.2	2 607	5 600	7/38, 4/36
30	10.0	100	78.5	104	3 287	7 000	3/35, 6/38
31	8.9	79.2	62.2	131	4 145	8 400	4/38, 7/40
32	8.0	64.0	50.3	162	5 230	10 500	—
33	7.1	50.4	39.6	206	6 591	13 100	—
34	6.3	39.7	31.2	261	8 311	16 800	
35	5.6	31.4	24.6	331	10 480	21 000	

Size AWG[1]	Diameter (mils)[2]	Area cmil[3]	Area in.² × 10⁻⁶	Ohms per 1000 ft.[4]	Feet per Pound[5]	Turns per in.² [6]	Typical Strandings[7]
36	5.0	25.0	19.6	415	13 210	26 000	—
37	4.5	20.2	15.9	512	16 660	31 000	—
38	4.0	16.0	12.6	648	21 010	39 000	—
39	3.5	12.2	9.62	847	26 500	53 000	—
40	3.1	9.61	7.55	1 080	34 400	65 000	—
41	2.8	7.84	6.16	1 320	42 000	—	—
42	2.5	6.25	4.91	1 660	53 000	—	—
43	2.2	4.84	3.80	2 140	68 000	—	—
44	2.0	4.00	3.14	2 590	82 000	—	—

1 Each wire size number increase represents a factor of 1.26 resistance increase over the previous size. Moving three gauge sizes higher doubles the resistance.

2 To convert to diameter in mm, multiply by 0.0254.

3 To convert table values to area in mm² × 10⁻⁶, multiply by 645.

4 To convert table values to Ω/m, multiply by 0.394.

5 To convert table values to m/kg, multiply by 0.672.

6 Turns per in.² cross-sectional area based on machine winding in a 1-in. × 1-in. channel. To convert to turns per cm², multiply by 0.155.

7 Other strandings often available. Note that total cross-sectional area of stranded wire may vary as much as +20%, –6% from solid-wire areas.

Color code for chassis wiring

Color	Conventional	Computer*
Black	Grounds, emitters	Grounds
Brown	Heaters, filaments	Control lines
Red	Positive HV, collector	Positive 5 V
Orange	Lower + supplies	Other + supplies
Yellow	Base, grid, neg dc, 5 V ac	Address, even No.
Green	6.3 V ac, filaments	Address, odd No.
Blue	Plates, anodes	Data, even No.
Violet	Misc.	Data, odd No.
Gray	AC power lines	I/O, even No.
White	Miscellaneous	I/O, odd No.

*Unofficial; recommended.

Popular coaxial connectors. *Plug* (male): center conductor is a pin; outer conductor usually fits over receptacle. *Jack* or *receptacle* (female): center conductor receives pin.

Description	BNC or UG number	UHF number
Line plug	88*, 260	PL-259
Line jack	89*, 261	
Chassis jack (4-screw-mount)	290, 447	SO-239
Chassis jack (1 hole & nut)	625, 912, 1094	
Double jack, line	914	PL-258
Double jack, chassis	414	UG-363
Double plug, line	491	
Tee; 2 jack, 1 plug	274	M-358

*For RG-58 size cable or smaller only.

Attenuation data for coax cables (next page) are for new cable at 20°C. Moisture intrusion, heat, and age can greatly increase attenuation.

Velocity factor for all coax cables in table is typically $0.7\ c$ to $0.8\ c$, or 225×10^6 m/s. Velocity factor for 300-Ω twin lead is about $0.8\ c$, and for open-wire line, about $0.97 c$.

Typical Characteristics of Popular Coaxial Cables

Type RGXX/U	Z (Ω)	O.D. (inches)	Weight (lb/100 ft)	V_{max} V, rms	Capacitance (pF/ft)	Attenuation (dB/100 ft) at f(MHz)						
						1	10	50	100	200	400	1000
8	52	0.40	11	5000	29.5	0.16	0.55	1.3	2.0	3.5	4.5	8.5
9	51	0.42	16	5000	30	0.12	0.47	1.4	1.9	2.9	4.4	8.0
11	75	0.41	10	5000	20.5	0.18	0.62	1.6	2.2	3.3	4.7	8.0
58	52	0.19	2.5	1900	30	0.40	1.3	3.2	5.0	8.0	12	22
59	73	0.24	3.2	2300	21	0.30	1.1	2.4	3.8	4.9	8.5	14
62	93	0.24	3.9	750	13.5	0.25	0.83	1.8	2.7	4.0	5.6	9.0
174	50	0.10	—	—	30	2.3	3.9	6.6	8.9	12	18	30
178	50	0.08	—	1000	29	2.6	5.6	10	14	20	28	46
179	75	0.11	1	1200	20	3.0	5.3	8.1	10	13	16	24
180	95	0.15	1.5	1500	15.5	2.4	3.3	4.6	5.7	7.6	11	17
213	50	0.41	12	5000	29.5	0.16	0.55	1.3	2.0	3.5	4.5	8.5
214	50	0.43	16	5000	30	0.12	0.47	1.4	1.9	2.9	4.4	8.0
223	50	0.22	3.6	1900	28.5	0.36	1.2	3.2	4.8	7.0	10	17

2.2 RESISTORS AND CAPACITORS

Memory aid for resistor color code:

Better	Be	Right	Or	Your	Great	Big	Venture	Goes	West
Blk	Brn	Red	Orn	Yel	Grn	Blu	Vio	Gry	Wht
0	**1**	**2**	**3**	**4**	**5**	**6**	**7**	**8**	**9**

Standard Values for 20%, 10%, and 5% tolerances

10% Values (20% values in bold)	5% Values (in addition to 10% values)
1.0	1.1
1.2	1.3
1.5	1.6
1.8	2.0
2.2	2.4
2.7	3.0
3.3	3.6
3.9	4.3
4.7	5.1
5.6	6.2
6.8	7.5
8.2	9.1
10	11

Standard Decade for 1%-tolerance components.

100	147	215	316	464	681
102	150	221	324	475	698
105	154	226	332	487	715
107	158	232	340	499	732
110	162	237	348	511	750
113	165	243	357	523	768
115	169	249	365	536	787
118	174	255	374	549	806
121	178	261	383	562	825
124	182	267	392	576	845
127	187	274	402	590	866
130	191	280	412	604	887
133	196	287	422	619	909
137	200	294	432	634	931
140	205	301	442	649	953
143	210	309	453	665	976

Color Codes for Resistors and Capacitors

Color	Significant Digit	Multiplier	Tolerance ΔR	Tolerance $C (\leq 10\ pF)$	Tolerance $C (> 10\ pF)$	Failure Rate /1000 hr (%)	Temp Coef (ppm/°C)	Working V, dc
Black	0	1	±20%	±2 pF	±20%	—	0	—
Brown	1	10	±1%	±0.1 pF	±1%	1.0	−33	100
Red	2	100	±2%	—	±2%	0.1	−75	—
Orange	3	1 000	±3%	±0.25 pF	±2.5%	0.01	−150	300
Yellow	4	10 000	GMV*	—	—	0.001	−220	—
Green	5	100 000	±5%	±0.5 pF	±5%	—	−330	500
Blue	6	1 000 000	—	—	—	—	−470	—
Violet	7	10 000 000	—	—	—	—	−750	—
Gray	8	0.01	—	±0.25 pF	—	—	+30	—
White	9	0.1	—	±1 pF	±10%	†	+500	—
Gold	—	0.1	±5%	—	±5%	—	+100	1000
Silver	—	0.01	±10%	—	±10%	—	Bypass	—
No Color	—	—	±20%	—	±20%	—	—	—

*Guaranteed Minimum Value: −0%, +100%. (Yel, col 4) †White 5th band indicates solderable terminal. (col 7)

64

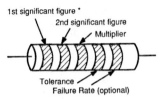

* Double-width band indicates wire-wound type.

Figure 2.1 Marking system for carbon-composition, carbon-film, and low-power wire-wound resistors.

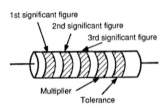

Figure 2.2 Markings for precision film resistors.

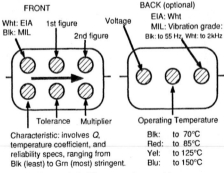

Figure 2.3 Mica capacitor markings.

The EIA code for ceramic capacitors specifies values in pF for whole numbers, or in µF for decimal numbers, followed by a letter indicating tolerance. Low-temperature limit, high-temperature limit, and maximum capacitance change over this temperature range are given, respectively, by a three-character code consisting of a letter, a number, and a letter. For temperature-compensating capacitors, this three-character code is replaced by the letter N (negative) or P (positive) and a number representing capacitance change in parts per million per degree Celsius. The letters NPO indicate zero temperature coefficient. The dc operating voltage limit may also be marked. Here are two examples.

0.001 K   Y5D   200 V
value tol. temp voltage

↓

0.001 µF, ±10%; ±3.3% change from −30 to +85°C.

68   J   N750
value tol. temp

↓

68 pF, ±5%; temp coef −750 ppm/°C

EIA Code for Ceramic Capacitors

Tolerance (%)	Temperature Characteristics		
	T_{min} (°C)		ΔC_{min} to C_{max} (%)
	X −55		A ±1
F ±1	Y −30		B ±1.5
G ±2	Z +10		C ±2.2
H ±3			D ±3.3
J ±5	T_{max} (°C)		E ±4.7
K ±10	5 +85		F ±7.5
M ±20	7 +125		P ±10
Z +80, −20			R ±15
P +100, −0 (also called			S ±22
GMV: Guaranteed			T +22, −33
Minimum Value)			U +22, −56
			V +22, −82

MIL-STD Resistor Designations: *Example*

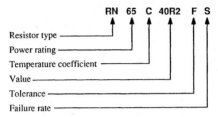

The resistor described above is a metal-film, 1/4-W type, with a temperature coefficient of ±50 ppm/°C, a value of 40.2 Ω, ±1%, with a failure certified to be not greater than 0.001% per 1000 hr under MIL specified test conditions. The codes on which this interpretation is based follow.

MIL STD Resistor Type and Power Rating Codes

- RA: variable, wirewound, precision

code:	20	25	30
power (W):	2	3	4

- RB: fixed, wirewound, precision

code:	08	16	17	18	19	52	53
power (W):	1/2	2/3	1	1 1/2	2	1	1/2

- RC: fixed, composition

code:	05	07	09	20	30	32	41	42
power (W):	1/8	1/4	1/2	1/2	1	1	2	2

- RD: power, film, noninductive

code:	31	33	35	37	39	60	65	70
power (W):	7	13	25	55	115	1	2	4

- RE: power, wirewound, with heat sink

code:	60	65	70	75	77	80
power (W):	7 1/2	20	25	50	100	250

- RL: fixed, film

code:	07	20	32	42
power (W):	1/4	1/2	1	2

- RN: fixed, film, high stability

code:	05	50	55	60	65	70	75
power (W):	$1/8$	$1/20$	$1/10$	$1/8$	$1/4$	$1/2$	1

- RP: variable, power, wirewound

code:	10	11	15	16	20	25	30
power (W):	25	12	50	25	75	100	150

- RV: variable, composition

code:	01	04	05	06
power (W):	$1/4$	2	$1/2$	$1/3$

- RW: fixed, power, wirewound

code:	55	56	67	68	69	70	74
power (W):	5	10	5	10	$2^1/2$	1	5

code:	78	79	80	81
power (W):	10	3	$2^1/4$	1

Note: The letter R following any of the two-letter codes shown above indicates that a reliability level has been established under the appropriate military specification.

Temperature coefficient: This position is used alternatively to specify temperature limit or lead structure:

J, E = ±25   H, C = ±50   K, O = ±100

all in parts per million per degree Celsius (ppm/°C).

Value: three (or four) digits. The first two (or three) are significant; the last one gives the number of zeros to be added. The letter R may be substituted for one of the numbers to indicate a decimal point, in which case all numbers following are significant figures, not multipliers.

Tolerance:

F = ±1%   G = ±2%   J = ±5%   K = ±10%

Failure rate:

M = 1%   P = 0.1%   R = 0.01%   S = 0.001%

all per 1000 hr under specified conditions.

Capacitance Calculations

For a single pair of plates, neglecting end effects, capacitance may be calculated by

$$C = 8.85 \times 10^{-14} \frac{KA}{d} \qquad \text{(metric)}$$

where C is in farads, A is the area between plates in cm^2, d is the plate spacing in cm, and K is the dielectric constant from the table below. Actual capacitance will be somewhat less, the error becoming significant as d becomes an appreciable fraction of the perimeter of A.

Dielectric Constants for Selected Materials

Material	K	Material	K
Vacuum	1.0000	Paper (paraffin-impregnated)	3—4
Air	1.0006	Rubber	3—5
Wood (dry)	1.5—5	Celluloid	4
Paper	2—3	Quartz	4—5
Oil	2—4	Formica	4.7
Teflon	2.1	Mica	4.5—8
Vaseline	2.16	Glass (Pyrex)	4.8
Polyethylene	2.3	Steatite (ceramic)	5—6
Lucite	2.5	Porcelain	5—6
Paraffin	2.5	Glass (window)	7—8
Wax	2.6	Water (distilled)	80
Polystyrene	2.6	Ceramic (high K)	to 7000
Plexiglass	2.8		

2.3 INDUCTORS AND TRANSFORMERS

Solenoid-wound coils (cylindrical shape, single layer or multilayer) have inductance given by

$$L_{\mu H} = \frac{d^2 N^2}{46d + 101l} \qquad \text{(diameter } d \text{ \& length } l \text{ in cm)}$$

To determine the number of turns required on a close-wound single-layer coil, given wire thickness t (cm), coil diameter d (cm), and required inductance L (µH):

$$N = \frac{\sqrt{(50.5Lt)^2 + 46d^3L} + 50.5Lt}{d^2}$$

69

Magnetic-core coils. Toroid, pot-core, C-I core, or E-I core have inductances given by

$$L_{\mu H} = 0.012 N^2 \frac{\mu A}{l_c} \qquad \text{(metric; no air gap)}$$

where N is the number of turns, A is the effective cross-sectional core area in cm^2, l_c is the magnetic path length in cm, and μ is the magnetic permeability of the core from the table below. If the core path contains an air gap, use

$$L_{\mu H} = 0.012 N^2 \frac{\mu A}{l_g + l_c/\mu} \qquad \text{(metric; air gap } l_g)$$

Permeability and Saturation of Magnetic Materials

Material	Relative Permeability (Air = 1.00)		B_{sat}
	μ_i (low level)	μ_m (high level)	(teslas)
Silicon iron	400	40 000	1.5
Iron/nickel alloy	3 000	20 000	1.0
Powdered iron (typ)	125	127	0.3
Ferrites 3B7, 3C8,			
TI, W-03	2 300	1 900	0.4
Ferrite 3E2A	5 000	1 800	0.3
Ferrite 3E3	12 000	1 900	0.4
Ferrite 3D3	750	1 500	0.3
Ferrite 4C4	125	600	0.3
Ferrite 1Z2	15	—	—

Because μ varies considerably with magnetizing force for most core materials, it is common practice to leave an air gap $l_g > l_c/\mu$ to linearize the inductor. For example, 200 turns on a 10-cm silicon-iron core with a 1-cm cross-section has an inductance of 19.2 mH at a low signal level and 1.92 H at a high signal level. If an air gap of 0.1 cm is introduced in the core, the inductor responds in a more nearly linear manner to different signal levels: 3.84 mH at low levels and 4.79 mH at high levels.

Maximum ac voltage across a magnetic-core coil:

$$V_{max\,(rms)} = 4.4 \times 10^{-4} f N A\, B_{sat}$$

where f is in Hz, N is number of turns, A is magnetic-core cross-sectional area in cm^2, and B_{sat} is in teslas.

Where inductance is known but N is not, this reduces to

$$V_{\text{max (rms)}} = 4.0\, B_{\text{sat}}\, f \sqrt{L\, l_e\, A}$$

where L is in henries and l_e is the effective magnetic path length in cm:

$$l_e = l_g + \frac{l_c}{\mu}$$

Maximum dc current in a magnetic-core coil:

$$I_{\text{sat}} = \frac{B_{\text{sat}}\, l_e}{1.26 \times 10^{-4}\, N}$$

where I is in amperes and the other terms are as defined above. Where L is known but N is not, use

$$I_{\text{sat}} = 0.87\, B_{\text{sat}} \sqrt{\frac{A\, l_c}{L}}$$

Pulse Transformers. When a dc voltage is applied across an inductor, the current rises to a final value V_P/R, as shown in Figure 2.4, below. However, the magnetic core may saturate before this current is reached, resulting in a sudden loss of inductor or transformer action. The product of source voltage and time to saturation is constant for any transformer and is called the *volt-microsecond* limit of a pulse transformer:

$$V_P\, t = I_{\text{sat}}\, L$$

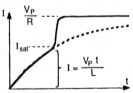

Figure 2.4 Pulse-transformer saturation.

Sine-wave ac across an inductor may also cause I_{sat} to be reached if the voltage is high enough and the frequency is low enough to permit this peak current to be reached. The ratio of rms voltage to frequency is constant for any transformer and is called the volt-per-hertz limit.

$$V_{\text{rms}} = 4.4\, I_{\text{sat}}\, L$$

Sometimes it is easier to test one of these limits than the other, making conversion to the other limit desirable:

$$V_{rms} = 4.4\, V_P\, t$$

Transformer and Audio Color Codes

Power transformers:

Black	Primary start
Black/yellow	Primary tap
Black/red	Primary finish
Red	High-voltage secondary
Red/yellow	High-voltage tap
Green	Low-voltage No. 1
Green/yellow	Low-voltage No. 1 tap
Brown	Low-voltage No. 2
Gray	Low-voltage No. 3
Yellow	Rectifier filament

Audio and IF transformers:

Blue	Primary signal source; finish
Red	Primary power feed; tap
Brown	Primary signal source; start
Green	Secondary signal out; finish
Black	Secondary ground return; tap
Yellow	Secondary signal out; start

Stereo phono leads:

Red	Right channel high
Green	Right channel low
White	Left channel high
Blue	Left channel low
Black	Ground

Stereo headphone plugs:

Tip	Right channel
Ring	Left channel
Barrel	Common

2.4 SEMICONDUCTOR DEVICES

Recommended Standard Semiconductor Types

Bipolar transistors:	NPN	PNP
Gen purpose plastic; 350 mW, 40 V	2N2222A	2N2907
	2N3904	2N3906
	2N4401	2N4403
Round metal TO-39 case; 5 W, 60 V	2N3053	2N4036
Power-tab TO-220 case; 40 W, 70 V	2N6292	2N6107
High-power TO-3 case; 115 W, 60 V	2N3055	2N5875
High-voltage, TO-39 case; 5 W, 250 V	2N3440	2N5416
VHF, med-power; TO-39, 5 W, 30 V	2N3866	
VHF, high-power; stud mount, 11 W	2N3375	

Power ratings are $T_{case} = 25\,°C$; voltages are $V_{B(CEO)}$

Other Semiconductors:	
N-channel JFET, plastic, 350 mW, 25 V	2N3819
Unijunction transistor, plastic, $I_{p(max)} = 2\ \mu A$	2N4948
Low-power SCR; plastic, 0.8 A, 200 V	2N5064
Higher-power SCR; TO-220, 16 A, 600 V	2N6404
Triac; flat case, 4 A, 400 V, $I_{GT} = 10$ mA	2N6073
Silicon signal diode; 20 mA, 75 V	1N914
Rectifier diode; plastic, 1 A, 400 V	1N4004
High-voltage rectifier; plastic, 1 A, 1 kV	1N4007
High-current rectifier; stud mount, 35 A, 100 V	1N1184
Fast-switching rectifier; plastic, 1 A, 200 V	1N4935

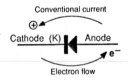

Figure 2.5 Diode symbol and current direction.

73

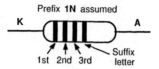

Figure 2.6 Diode color-band marking system.

Color	Number	Suffix		Color	Number	Suffix
Black	0	none		Green	5	E
Brown	1	A		Blue	6	F
Red	2	B		Violet	7	G
Orange	3	C		Gray	8	H
Yellow	4	D		White	9	J

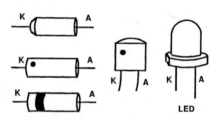

Figure 2.7 Diode lead identification.

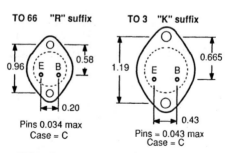

Figure 2.8 Popular transistor case dimensions.

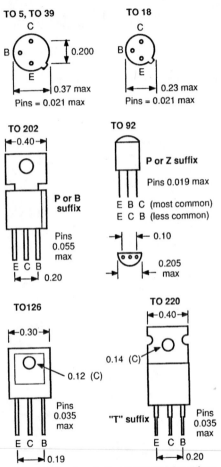

Figure 2.8 (continued) Transistor case dimensions.

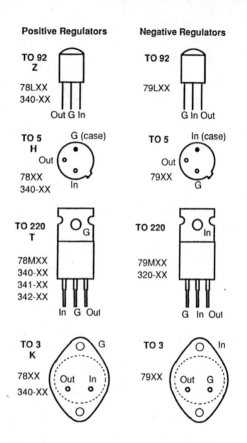

Figure 2.9 Three-terminal regulator pinouts. Front views of TO 92 and TO 220; bottom views of TO 5 and TO 3 cases are shown.

Package Pin-outs for Popular ICs
* indicates negative-true (low-active) pin.

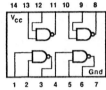

Similar pin-outs for:
7400 quad 2-in NAND
7403 open collector NAND
7408 quad 2-in AND
7409 open collector AND
7426 high-voltage NAND
7432 quad 2-in OR
7437 invert NAND buffer
7438 inv NAND, O. C.
7486 quad exclusive OR
74132 NAND schmitt trig

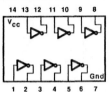

Similar pin-outs for:
7404 hex inverter
7405 hex inv, O. C.
7406 inv buf, O. C., H. V.
7407 non-inv, O. C., H.V.
7414 hex inv schmitt trig
7416 hex inv, O. C., H. V.
7417 non-inv, O. C., H. V.

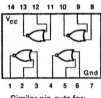

Similar pin-outs for:
7402 quad 2-input NOR
7401 quad 2-input NAND
7428 quad NOR buffers
7433 NOR buffer, O. C.
74128 NOR line drivers

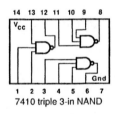

7410 triple 3-in NAND

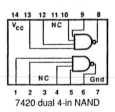

7420 dual 4-in NAND

77

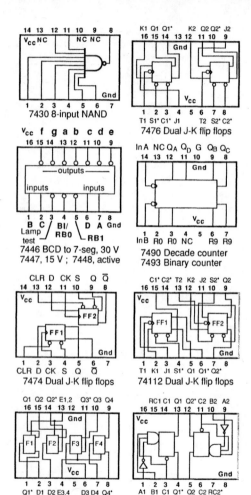

7430 8-input NAND

14 13 12 11 10 9 8
Vcc NC NC NC
Gnd
1 2 3 4 5 6 7

7476 Dual J-K flip flops

K1 Q1 Q1* K2 Q2 Q2* J2
16 15 14 13 12 11 10 9
Gnd
Vcc
1 2 3 4 5 6 7 8
T1 S1* C1* J1 T2 S2* C2*

7446 BCD to 7-seg, 30 V
7447, 15 V ; 7448, active

Vcc f g a b c d e
16 15 14 13 12 11 10 9
— outputs —
inputs   inputs
1 2 3 4 5 6 7 8
B C Bl/ D A Gnd
Lamp RB0 RB1
test

7490 Decade counter
7493 Binary counter

In A NC QA QD G QB QC
14 13 12 11 10 9 8
Gnd
Vcc
1 2 3 4 5 6 7
In B R0 R0 NC R9 R9

7474 Dual J-K flip flops

CLR D CK S Q Q̄
14 13 12 11 10 9 8
Vcc
FF2
FF1
Gnd
1 2 3 4 5 6 7
CLR D CK S Q Q̄

74112 Dual J-K flip flops

C1* C2* T2 K2 J2 S2* Q2*
16 15 14 13 12 11 10 9
Vcc
FF1   FF2
Gnd
1 2 3 4 5 6 7 8
T1 K1 J1 S1* Q1 Q1* Q2*

7475 Quad latch

Q1 Q2 Q2* E1,2 Q3* Q3 Q4
16 15 14 13 12 11 10 9
Gnd
F1 F2 F3 F4
Vcc
1 2 3 4 5 6 7 8
Q1* D1 D2 E3,4 D3 D4 Q4*

74123 Dual one-shots

RC1 C1 Q1 Q2* C2 B2 A2
16 15 14 13 12 11 10 9
Vcc
Gnd
1 2 3 4 5 6 7 8
A1 B1 C1 Q1* Q2 Q2 RC2*

78

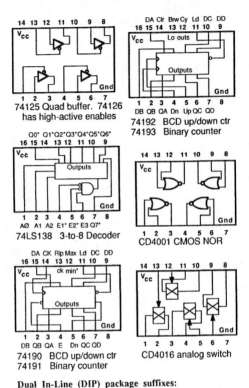

74125 Quad buffer. 74126 has high-active enables

74192 BCD up/down ctr
74193 Binary counter

74LS138 3-to-8 Decoder

CD4001 CMOS NOR

74190 BCD up/down ctr
74191 Binary counter

CD4016 analog switch

Dual In-Line (DIP) package suffixes:

N = plastic, J = ceramic, T and W = flat packs

TTL Logic Levels for 54, 74, LS, S, and ALS gates.

Valid high input = 2.0 V and above.
Valid low input = 0.7 V and below. †
Guaranteed high output = 2.4 V or above.
Guaranteed low output = 0.5 V or below.

  † Some schmitt triggers require 0.5 V or below.

74LS138 Decoder

1	A	Vcc	16
2	B	Y0	15
3	C	Y1	14
4	$\overline{CS}$	Y2	13
5	$\overline{CS}$	Y3	12
6	CS	Y4	11
7	Y7	Y5	10
8	Gnd	Y6	9

74LS244 Buffer

1	$\overline{CE}$	Vcc	20
2	I0	$\overline{CE}$	19
3	D7	D0	18
4	I1	I7	17
5	D6	D1	16
6	I2	I6	15
7	D5	D2	14
8	I3	I5	13
9	D4	D3	12
10	Gnd	I4	11

74LS373 Latch

1	$\overline{OE}$	Vcc	20
2	Q0	Q7	19
3	D0	D7	18
4	D1	D6	17
5	Q1	Q6	16
6	Q2	Q5	15
7	D2	D5	14
8	D3	D4	13
9	Q3	Q4	12
10	Gnd	CE	11

386 AF amp

1	Av	Av	8
2	-In	C	7
3	+In	Vcc	6
4	Gnd	Out	5

339 Comparator

1	Y2	Y3	14
2	Y1	Y4	13
3	Vcc	Gnd	12
4	1-	4+	11
5	1+	4-	10
6	2-	3+	9
7	2+	3-	8

741 Op amp

1	Nul	NC	8
2	-In	+V	7
3	+In	Out	6
4	-V	Nul	5

555 Timer

1	Gnd	Vcc	8
2	Trig	Dis	7
3	Out	Thr	6
4	Rst	Ctrl	5

565 PLL

1	-V	NC	14
2	-In	NC	13
3	+In	NC	12
4	VCO out	NC	11
5	Ø in	+V	10
6	Ref	C	9
7	VCO in	R	8

723 Regulator

1	NC	NC	14
2	Lim	Com	13
3	Sen+	Sup	12
4	-In	Vc	11
5	+In	Vo	10
6	Ref	Vz	9
7	-Sup	NC	8

7-seg display

1	a	Com	14
2	f	b	13
3	Com	NC	12
4	NC	g	11
5	NC	c	10
6	DP	Com	9
7	e	d	8

Microprocessor Stick-on Labels

	8080	
1	A10	A11 40
2	GND	A14 39
3	D4	A13 38
4	D5	A12 37
5	D6	A15 36
6	D7	A9 35
7	D3	A8 34
8	D2	A7 33
9	D1	A6 32
10	D0	A5 31
11	−5	A4 30
12	RST	A3 29
13	HLD	+12 28
14	INT	A2 27
15	Ø2	A1 26
16	INE	A0 25
17	DIN	WAI 24
18	$\overline{WR}$	RDY 23
19	SYN	Ø1 22
20	+5	HDA 21

	8085	
1	XT1	Vcc 40
2	XT2	HLD 39
3	RST	HDA 38
4	SOD	CLK 37
5	SID	$\overline{RST}$ 36
6	TRP	RDY 35
7	RS7	IO/$\overline{M}$ 34
8	RS6	S1 33
9	RS5	$\overline{RD}$ 32
10	INT	$\overline{WR}$ 31
11	$\overline{IAK}$	ALE 30
12	AD0	S0 29
13	AD1	A15 28
14	AD2	A14 27
15	AD3	A13 26
16	AD4	A12 25
17	AD5	A11 24
18	AD6	A10 23
19	AD7	A9 22
20	Vss	A8 21

	Z80	
1	A11	A10 40
2	A12	A9 39
3	A13	A8 38
4	A14	A7 37
5	A15	A6 36
6	CK	A5 35
7	D4	A4 34
8	D3	A3 33
9	D5	A2 32
10	D6	A1 31
11	Vcc	A0 30
12	D2	GND 29
13	D7	$\overline{RF}$ 28
14	D0	$\overline{M1}$ 27
15	D1	$\overline{RST}$ 26
16	$\overline{INT}$	$\overline{BR}$ 25
17	$\overline{NMI}$	$\overline{WA}$ 24
18	$\overline{HLT}$	BAK 23
19	$\overline{MR}$	$\overline{WR}$ 22
20	$\overline{IO}$	$\overline{RD}$ 21

	8086	
1	GND	Vcc 40
2	A14	A15 39
3	A13	A16 38
4	A12	A17 37
5	A11	A18 36
6	A10	A19 35
7	A9	$\overline{BHE}$ 34
8	A8	$\overline{MAX}$ 33
9	A7	$\overline{RD}$ 32
10	A6	HLD 31
11	A5	HDA 30
12	A4	$\overline{WR}$ 29
13	A3	M/$\overline{IO}$ 28
14	A2	DT/$\overline{R}$ 27
15	A1	$\overline{DEN}$ 26
16	A0	ALE 25
17	NMI	$\overline{INA}$ 24
18	INT	$\overline{TST}$ 23
19	CK	RDY 22
20	Vss	RST 21

68000

Pin	Signal	Signal	Pin
1	D4	D5	64
2	D3	D6	63
3	D2	D7	62
4	D1	D8	61
5	D0	D9	60
6	$\overline{AS}$	D10	59
7	$\overline{UDS}$	D11	58
8	$\overline{LDS}$	D12	57
9	R/$\overline{W}$	D13	56
10	$\overline{DTAK}$	D14	55
11	$\overline{BG}$	D15	54
12	$\overline{BGA}$	GND	53
13	$\overline{BR}$	A23	52
14	Vcc	A22	51
15	CLK	A21	50
16	GND	Vcc	49
17	$\overline{HLT}$	A20	48
18	$\overline{RST}$	A19	47
19	$\overline{VMA}$	A18	46
20	E	A17	45
21	$\overline{VPA}$	A16	44
22	$\overline{BER}$	A15	43
23	$\overline{IP2}$	A14	42
24	$\overline{IP1}$	A13	41
25	$\overline{IP0}$	A12	40
26	FC2	A11	39
27	FC1	A10	38
28	FC0	A9	37
29	A1	A8	36
30	A2	A7	35
31	A3	A6	34
32	A4	A5	33

27128

Pin	Signal	Signal	Pin
1	Vpp	Vcc	28
2	A12	PGM	27
3	A7	A13	26
4	A6	A8	25
5	A5	A9	24
6	A4	A11	23
7	A3	$\overline{OE}$	22
8	A2	A10	21
9	A1	$\overline{CE}$	20
10	A0	D7	19
11	D0	D6	18
12	D1	D5	17
13	D2	D4	16
14	GND	D3	15

2716

Pin	Signal	Signal	Pin
1	A7	Vcc	24
2	A6	A8	23
3	A5	A9	22
4	A4	Vpp	21
5	A3	$\overline{OE}$	20
6	A2	A10	19
7	A1	$\overline{CE}$	18
8	A0	D7	17
9	D0	D6	16
10	D1	D5	15
11	D2	D4	14
12	Vss	D3	13

6800

Pin	Signal		Signal	Pin
1	GND		$\overline{RST}$	40
2	$\overline{HLT}$		TSC	39
3	Ø1		NC	38
4	$\overline{IRQ}$		Ø2	37
5	VMA		DBE	36
6	$\overline{NMI}$		NC	35
7	BA		R/$\overline{W}$	34
8	Vcc		D0	33
9	A0		D1	32
10	A1		D2	31
11	A2		D3	30
12	A3		D4	29
13	A4		D5	28
14	A5		D6	27
15	A6		D7	26
16	A7		A15	25
17	A8		A14	24
18	A9		A13	23
19	A10		A12	22
20	A11		Gnd	21

6800

6802

Pin	Signal		Signal	Pin
1	GND		$\overline{RST}$	40
2	$\overline{HLT}$		E X	39
3	MR		XTL	38
4	$\overline{IRQ}$		E	37
5	VMA		RE	36
6	$\overline{NMI}$		VR	35
7	BA		R/$\overline{W}$	34
8	Vcc		D0	33
9	A0		D1	32
10	A1		D2	31
11	A2		D3	30
12	A3		D4	29
13	A4		D5	28
14	A5		D6	27
15	A6		D7	26
16	A7		A15	25
17	A8		A14	24
18	A9		A13	23
19	A10		A12	22
20	A11		Gnd	21

6802

6809

Pin	Signal		Signal	Pin
1	GND		$\overline{HLT}$	40
2	$\overline{NMI}$		XTL	39
3	$\overline{IRQ}$		EXT	38
4	$\overline{FIR}$		$\overline{RST}$	37
5	BS		MR	36
6	BA		Q	35
7	Vcc		E	34
8	AØ		$\overline{BRQ}$	33
9	A1		R/$\overline{W}$	32
10	A2		DØ	31
11	A3		D1	30
12	A4		D2	29
13	A5		D3	28
14	A6		D4	27
15	A7		D5	26
16	A8		D6	25
17	A9		D7	24
18	A10		A15	23
19	A11		A14	22
20	A12		A13	21

6809

6502

Pin	Signal		Signal	Pin
1	GND		$\overline{RST}$	40
2	RDY		Ø2	39
3	Ø1		SO	38
4	$\overline{IRQ}$		Ø0	37
5	NC		NC	36
6	$\overline{NMI}$		NC	35
7	SYN		R/W	34
8	Vcc		D0	33
9	A0		D1	32
10	A1		D2	31
11	A2		D3	30
12	A3		D4	29
13	A4		D5	28
14	A5		D6	27
15	A6		D7	26
16	A7		A15	25
17	A8		A14	24
18	A9		A13	23
19	A10		A12	22
20	A11		GND	21

6502

2.5 MISCELLANEOUS COMPONENTS

Pilot-Lamp Data
Popular types in Boldface.

Lamp	Volts	mA	Base
PR-2	2.4	500	Flange
PR-3	3.6	500	Flange
PR-4	2.3	270	Flange
PR-6	2.5	300	Flange
PR-7	3.7	300	Flange
PR-12	5.95	500	Flange
PR-13	4.75	500	Flange
PR-18	6.3	150	Flange
12	6.3	150	2-pin
13	3.8	300	Screw
14	2.5	300	Screw
19	14.4	100	Pin
40	6.3	150	Screw
43	2.5	500	Bayonet
44	6.3	250	Bayonet
45	3.2	350	Bayonet
46	6.3	250	Screw
47	6.3	150	Bayonet
48	2.0	60	Screw
49	2.0	60	Bayonet
50	6.3	200	Screw
51	7.5	220	Bayonet
53	14.4	120	Bayonet
55	7.0	410	Bayonet
57	14.0	240	Bayonet
73	14.0	80	Wedge
85	28.0	40	Wedge
86	6.3	200	Wedge
88	6.8	1900	2-btn bay
93	12.8	1000	Bayonet
112	1.2	220	Screw
161	14.0	190	Wedge
194	14.0	270	Wedge

Lamp	Volts	mA	Base
219	6.3	250	Bayonet
222	2.2	250	Screw
233	2.3	270	Screw
239	6.3	360	Bayonet
259	6.3	250	Wedge
307	28.0	660	Bayonet
313	28.0	170	Bayonet
327	28.0	40	Flange
328	6.0	200	Flange
330	14.0	80	Flange
334	28.0	40	Groove
335	28.0	40	Screw
336	14.0	80	Groove
338	2.7	60	Flange
345	6.0	40	Flange
356	28.0	170	Bayonet
370	18.0	40	Flange
376	28.0	60	Flange
381	6.3	200	Flange
382	14.0	80	Flange
385	28.0	40	Flange
386	14.0	80	Groove
387	28.0	40	Flange
388	28.0	40	Groove
394	12.0	40	Flange
398	6.3	200	Groove
399	28.0	170	Screw
656	28.0	60	Wedge
680	5.0	60	Wire
682	5.0	60	Flange
685	5.0	60	Flange
715	5.0	115	Wire
755	6.3	150	Bayonet
756	14.0	80	Bayonet
757	28.0	80	Bayonet
1156	12.8	2100	Bayonet
1157	12.8	2100	Bayonet

Lamp	Volts	mA	Base
1251	28.0	230	Bayonet
1450	24.0	35	Bayonet
1490	3.2	160	Bayonet
1493	6.5	2750	2-con bay
1495	28.0	300	Bayonet
1813	14.4	100	Bayonet
1815	14.0	200	Bayonet
1816	13.0	330	Bayonet
1818	24.0	170	Bayonet
1819	28.0	40	Min. bay.
1820	28.0	100	Bayonet
1821	28.0	170	Screw
1822	36.0	100	Bayonet
1828	37.5	50	Bayonet
1829	28.0	70	Bayonet
1835	55.0	50	Bayonet
1847	6.3	150	Min. bay.
1864	28.0	170	Bayonet
1866	6.3	250	Bayonet
1891	14.0	240	Min. bay.
1892	14.4	120	Min. bay.
1893	14.0	330	Bayonet
1895	14.0	270	Bayonet
2232	28.0	640	Bayonet
3150	5.0	60	Flange
6833	5.0	60	Wire
6838	28.0	24	Wire
6839	28.0	24	Flange
7152	5.0	115	Wire
7153	5.0	115	Wire
7219	12.0	60	Wire
7265	5.0	60	Bi-pin
7327	28.0	40	Bi-pin
7328	6.0	200	Bi-pin
7333	5.0	60	Flange
7335	5.0	115	Flange

Lamp	Volts	mA	Base
7341	28.0	65	Flange
7361	5.0	60	Bi-pin
7371	12.0	40	Bi-pin
7373	14.0	100	Bi-pin
7374	28.0	40	Bi-pin
7381	6.3	200	Bi-pin
7382	14.0	80	Bi-pin
7387	28.0	40	Bi-pin
7632	28.0	40	Bi-pin
7680	5.0	60	Bi-pin
7683	5.0	60	Bi-pin
7715	5.0	115	Bi-pin
7876	28.0	60	Bi-pin

Neon Lamp	R (kΩ)	mA	Base
A1A (NE-2)	100	0.7	Wire leads
A1B	100	0.3	Wire leads
A1C	47	1.2	Wire leads
A2A	100	0.7	Wire leads
A2B (NE-2V)	100	0.7	Wire leads
A9A (NE-2E)	100	0.7	Wire leads
B1A (NE-51)	220	0.3	Bayonet
B2A (NE-51H)	47	1.2	Bayonet
B5A (NE-17)	30	2.0	2-con bay
B6A (NE-21)	30	2.0	Bayonet
B7A (NE-45)	30	2.0	Screw
B8A (NE-48)	30	2.0	Bayonet
B9A (NE-48)	30	2.0	2-con bay
C2A (NE-2H)	30	1.9	Wire leads
C7A (NE-2D)	100	0.7	Flange
C9A (NE-2J)	30	1.9	Flange
J5A (NE-30)	7.5	8.0	Screw
K1A5 (NE-84)	30	1.9	Wedge
K1B1	47	1.2	Wedge
K1C5	100	0.7	Wedge
R1A	7.5	8.0	2-con bay

Efficiency of various light sources
Light energy out ÷ electrical energy in.

Light-emitting diode	1%
Neon glow lamp	1%
Incandescent lamp	1.5 — 3%
Mercury vapor lamp	6%
Fluorescent lamp	7 — 12%
High-pressure sodium lamp	16 — 22%

Thermocouple characteristics

Type J. Iron-constantan.
Useful to 750 °C. Output approximately 55 μV/ °C.
Colors: White = iron (+); Red = constantan (–)

Type K. Chromel-alumel. To 1250 °; $V_0 \approx 39$ μV/ °C.
Colors: Yellow = chromel (+); Red = alumel (–)

Type T. Copper-constn. To 350 °C; $V_0 \approx 51$ μV/ °C.
Colors: Blue = copper (+); Red = constantan (–)

Type E. Chromel-constn. To 900 °C; $V_0 \approx 72$ μV/ °C.
Colors: Violet = chromel (+); Red = constantan (–)

Type R. Platinum w/ 13% rhodium-platinum.
Useful to 1450 °C; Output approximately 11 μV/ °C.
Colors: Black = Pt/13% Rh (+); Red = Pt (–)

Type S. Platinum w/ 10% rhodium-platinum.
Useful to 1450 °C; Output approximately 10 μV/ °C.
Colors: Black = Pt/10% Rh (+); Red = Pt (–)

Type B. Platinum w/ 30% rhodium-Pt w/ 6% Rh.
Useful to 1700 °C; Output approximately 7.3 μV/ °C.
Colors: Gray = Pt/30% Rh (+); Red = Pt/6% Rh (–)

Type G. Tungsten-tungsten w/ 26% rhenium.
Useful to 2320 °C; Output approximately 17 μV/ °C.
Colors: White = tungsten (+); Red = Tng-26% Rh (–)

Battery Cross Reference

Common Designation	AAA	AA	C	D	No. 6	Lantern	Transistor
Voltage (V)	1.5	1.5	1.5	1.5	1.5	6	9
Height (inches)	1.69	1.88	1.81	2.25	6.0	4.37	1.93
Width or dia. (inches)	0.39	0.53	0.94	1.25	2.50	2.63	1.03
Depth (inches)	—	—	—	—	—	2.63	0.69
NEDA number*	24	15	14	13	905	908	1604
IEC number	R-03	R-6	R-14	R-20	R-40	4R23	6F22
Eveready (alkaline)	912 (E 92)	915 (E 91)	935 (E 93)	950 (E 95)	156	509	216, 522BP
Burgess	7	Z	1	2	6	F4M	2U6
Mallory	MN2400	MN1500	MN1400	MN1300	MN905	MN908	MN1604
RCA	VS074	VS034	VS035	VS036	VS0065	VS040C	VS323
Mercury (1.4 V/cell)	—	xx9, x502	—	xx42	—	—	146x

* Suffix A indicates alkaline.

Relay Contact Designations

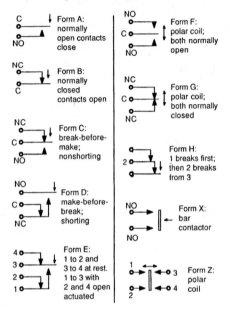

C / NO — Form A: normally open contacts close

NC / C — Form B: normally closed contacts open

NC / C / NO — Form C: break-before-make; nonshorting

NO / C / NC — Form D: make-before-break; shorting

4 / 3 / 2 / 1 — Form E: 1 to 2 and 3 to 4 at rest. 1 to 3 with 2 and 4 open actuated

NO / C / NO — Form F: polar coil; both normally open

NC / C / NC — Form G: polar coil; both normally closed

2 — Form H: 1 breaks first; then 2 breaks from 3

NO / NO — Form X: bar contactor

1 / 2 → 3 / 4 — Form Z: polar coil

90

3

Circuit Analysis and Design

3.1 RESISTIVE CIRCUITS

Dropping-resistor calculation. In the circuit of Figure 3.1 with source V_S and load R_L, find a dropping resistance R_D that drops the source to the required load voltage V_{RL} (less than V_S). A disadvantage is that variations in load current cause changes in load voltage.

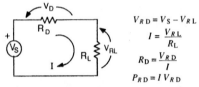

$$V_{RD} = V_S - V_{RL}$$

$$I = \frac{V_{RL}}{R_L}$$

$$R_D = \frac{V_{RD}}{I}$$

$$P_{RD} = I V_{RD}$$

Figure 3.1 Calculations for a dropping resistor.

Loaded voltage divider: Given Figure 3.2 with source V_S, load R_L, required voltage V_O, and regulation factor η by which V_O may rise when R_L is removed, find dividing resistors R_1 and R_2. Decreasing R_1 and R_2 improves regulation but causes wasteful current drain from V_S.

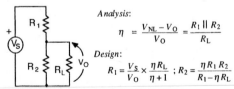

Analysis:

$$\eta = \frac{V_{NL} - V_O}{V_O} = \frac{R_1 \,\|\, R_2}{R_L}$$

Design:

$$R_1 = \frac{V_S}{V_O} \times \frac{\eta R_L}{\eta + 1} \quad ; \quad R_2 = \frac{\eta R_1 R_2}{R_1 - \eta R_L}$$

Figure 3.2 Calculations for a loaded voltage divider.

Variable-resistance network. Determine the values of R_1, R_v, and R_2 in Figure 3.3, such that a required minimum resistance R_{low} is reached when R_v is set to zero resistance, and a required maximum R_{high} is reached with R_v adjusted to its maximum resistance. Note that potentiometers are commonly available only in values following a 1—2—5 sequence, and R_v must be limited to this series.

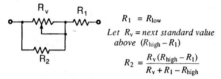

$$R_1 = R_{low}$$

Let R_v = next standard value above $(R_{high} - R_1)$

$$R_2 = \frac{R_v(R_{high} - R_1)}{R_v + R_1 - R_{high}}$$

Figure 3.3 Obtaining desired R_{max} and R_{min}.

Variable voltage divider, unloaded (Figure 3.4, below): Given two voltage sources V_1 and V_2, and a bleeder current I, find values of R_1, R_v, and R_2 that will produce the required output-voltage range V_{max} to V_{min}. Often V_2 is zero (ground). Any of the voltages may be negative, in which case they are treated algebraically in the formulas. One of the three resistances, usually R_v, will have to be chosen from source and load current considerations to begin the calculations. If there is a load current I_L, accuracy will be impaired by a maximum amount I_L / I_{bias}, where I_{bias} is the current through R_1, R_v, and R_2.

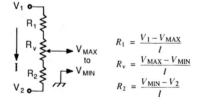

$$R_1 = \frac{V_1 - V_{MAX}}{I}$$

$$R_v = \frac{V_{MAX} - V_{MIN}}{I}$$

$$R_2 = \frac{V_{MIN} - V_2}{I}$$

Figure 3.4 Designing a variable voltage divider for maximum and minimum outputs. Treat negative voltage values algebraically.

3.2 NETWORK THEOREMS

Thévenin's theorem states that a two-terminal network consisting of any number of resistors and voltage sources can be replaced by an equivalent circuit containing one voltage source and one resistance. The two terminals selected are usually (but not always) those across which the output voltage appears. To form the equivalent:

1. Remove a circuit element between the two terminals.

2. Determine the voltage between the two terminals with the element removed. This is V_{Th}.

3. Mentally replace all voltage sources with a short circuit and calculate the resistance of the remaining network between the two terminals. The output element is still removed. The resistance calculated is R_{Th}.

4. Replace the original network with V_{Th} and R_{Th} in series. Replace the output element and analyze the resulting simple circuit for V_O.

Figure 3.5 and the following calculations show how Thévenin's theorem is used to solve a loaded R-C time-constant problem. The load element removed is C.

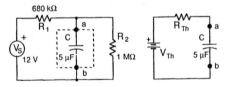

Figure 3.5 Thévenin equivalent of an R-C circuit.

$$V_{ab} = V_{Th} = V_S \frac{R_2}{R_1 + R_2} = 12 \frac{1}{1 + 0.68} = 7.14 \text{ V}$$

$$R_{Th} = R_1 \| R_2 = 0.405 \text{ M}\Omega$$

$$V_{C(max)} = V_{Th} = 7.14 \text{ V}$$

$$\tau = R_{Th}C = 0.405 \text{ M}\Omega \times 5 \text{ μF} = \textbf{2.0 s}$$

Current sources are idealizations presumed to deliver a fixed output current regardless of the load impedance connected across them. Voltage sources are presumed to deliver a constant voltage regardless of load current. Real voltage sources have a small resistance R_{Th} in series with the ideal voltage source. Real current sources have a large resistance R_N in parallel with the ideal current source. Figure 3.6, below, shows how a real voltage source can be converted to an equivalent real current source, and vice versa.

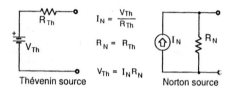

$$I_N = \frac{V_{Th}}{R_{Th}}$$

$$R_N = R_{Th}$$

$$V_{Th} = I_N R_N$$

Thévenin source   Norton source

Figure 3.6 Thévenin-Norton conversions.

Norton's theorem is similar to Thévenin's. It produces a current-source equivalent $I_N \| R_N$ as Thévenin produces a voltage-source equivalent. To form a Norton equivalent circuit:

1. Remove the output element from the selected output terminals.

2. Calculate the current through a short circuit placed across these terminals. This is I_N.

3. Mentally open-circuit all current sources and short-circuit all voltage sources. Calculate R_N between the output terminals.

4. Replace the original network with $I_N \| R_N$.

Millman's theorem permits any number of parallel voltage sources V_n (each with its series resistance R_n) to be represented as a single voltage source V_{Th} and series resistance R_{Th}. Figure 3.7 and the equations on the next page illustrate a Millman simplification.

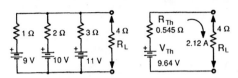

Figure 3.7 Simplification by Millman's theorem.

$$R_{Th} = R_1 \| R_2 \| R_3 \cdots = \cfrac{1}{\cfrac{1}{R_1} + \cfrac{1}{R_2} + \cfrac{1}{R_3} + \cdots}$$

$$V_{Th} = R_{Th} \left(\frac{V_1}{R_1} + \frac{V_2}{R_2} + \frac{V_3}{R_3} + \cdots \right)$$

The superposition theorem permits solution of many networks containing multiple sources with elementary techniques. It states that the current (or voltage) response in each element of a linear network is equal to the algebraic sum of the responses produced by each source acting independently. The restriction to "linear" elements precludes devices such as thermistors and saturating inductors. Unused voltage sources are shorted and unused current sources are opened at each step of the analysis. In Figure 3.8 and the calculations below, the second source is a zener diode. The zener may be treated as a voltage source if it is assured that it is always conducting.

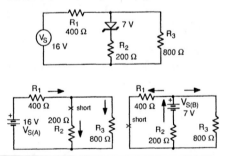

Figure 3.8 Circuit with two voltage "sources," above, and response to each source, below.

$$I_{R1A} = \frac{V_{SA}}{R_1 + R_2 \| R_3} = \frac{16}{400 + 160} = 0.0286 \text{ A}$$

$$V_{R1A} = I_{R1A} R_1 = 0.0286 \times 400 = 11.44 \text{ V}$$

$$V_{R3A} = V_{SA} - V_{R1A} = 16 - 11.44 = 4.57 \text{ V}$$

$$I_{R2B} = \frac{V_{SB}}{R_2 + R_1 \| R_3} = \frac{7}{200 + 267} = 0.0150 \text{ A}$$

$$V_{R2B} = I_{R2B} R_2 = 0.0150 \times 200 = 3.00 \text{ V}$$

$$V_{R3B} = V_{SB} - V_{R2B} = 7.00 - 3.00 = 4.00 \text{ V}$$

$$V_{R3} = V_{R3A} + V_{R3B} = 4.57 + 4.00 = 8.57 \text{ V}$$

The reciprocity theorem applies only to single-source networks. It states that the current response in any line A caused by a voltage source in any line B will be the same if the source is moved to line A and the current is measured in line B. Figure 3.9 and the calculations below show how the theorem is used to find the time constant and final current in a complex R-L circuit.

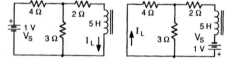

Figure 3.9 Reciprocity interchanges the source and the current response.

$$\tau = \frac{L}{R_2 + R_3 \| R_4} = \frac{5}{2 + 1.71} = 1.35 \text{ s}$$

$I_{L(\text{final})} = I_{R4}$ with L a short circuit

$$= \frac{3}{7} I_{\text{total}} = \frac{3}{7} \frac{1}{2 \, \Omega + 3 \Omega \| 4 \Omega} = 0.115 \text{ A}$$

Delta-wye conversion of part of a circuit will often permit a series combination that was not possible before, leading to the solution of the circuit. Figure 3.10, below, shows the delta and wye configurations and notation. The Δ and Y configurations are sometimes termed π and T, respectively.

$$R_1 = \frac{R_A R_C}{R_A + R_B + R_C} \qquad R_2 = \frac{R_B R_C}{R_A + R_B + R_C}$$

$$R_3 = \frac{R_A R_B}{R_A + R_B + R_C}$$

If $R_A = R_C = R_C$ (call them R_Δ), then $R_1 = R_2 = R_3$ (call them R_Y), and the above equations reduce to

$$R_Y = \frac{R_\Delta}{3}$$

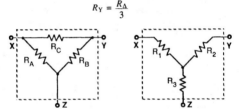

Figure 3.10 Delta and wye equivalent circuits.

Wye-delta conversion of a part of a circuit may permit new parallel combinations, leading to the solution of the circuit.

$$R_A = \frac{R_1 R_2 + R_1 R_3 + R_2 R_3}{R_2}$$

$$R_B = \frac{R_1 R_2 + R_1 R_3 + R_2 R_3}{R_1}$$

$$R_C = \frac{R_1 R_2 + R_1 R_3 + R_2 R_3}{R_3}$$

Where all Y-resistors are equal, all Δ resistors will be equal:

$$R_\Delta = 3R_Y$$

3.3 GENERAL CIRCUIT-ANALYSIS TECHNIQUES

Mesh analysis, also called loop analysis, is capable of solving any linear resistive circuit. Extensions of the technique permit solution of any resistive-reactive circuit. The method is based upon Kirchhoff's voltage law: *The sum of the voltage drops (resistive) is equal to the sum of the voltage rises (sources) for every closed path around a circuit.* To solve a circuit by mesh analysis:

1. Draw the circuit in a planer form (no wires crossing) if possible. Convert any current sources to voltage sources.

2. Draw clockwise current-indicating arrows within each loop, marking them I_1, I_2, and so on. Figure 3-11 on the next page shows an example. Conventional current (positive-to-negative) is indicated, but electron flow could be used. The only requirement is consistency.

3. Set up a table for the equations that will be obtained from the circuit. The number of rows equals the number of loops. There are three columns: A and B to the left of the equals signs and C to the right. See page 99.

4. For each equation N, the column-A term is the loop current I_N times the sum of the resistances through which I_N passes.

5. The column-B term is subtracted from the column-A term. For each equation N it consists of any resistance(s) that carry another current besides I_N, times that other current I_X. It is possible that this mutual resistor may carry more than one other current besides I_N, in which case the column-B term will have the form $-R_4(I_5 + I_6)$. It is also possible that loop N will have two or more resistors carrying current from two or more other loops, in which case there will be two or more column-B terms, for example: $-R_7I_8 - R_9I_{10}$.

6. The column-C term is the algebraic sum of the voltage sources through which I_N passes. A source is positive if it is acting in the direction indicated by I_N and negative if it acts against I_N.

7. Solve the resulting simultaneous equations for each I_N by substitution or by determinants (see pages 53 and 54). Negative values for I simply mean that the current for that loop is actually counterclockwise. Figure 3.11, below, shows presumed loop currents and actual branch currents, respectively, for the example problem.

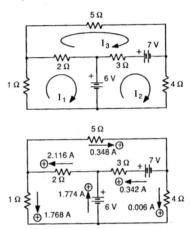

Figure 3.11 Mesh analysis: presumed loop currents (top) and actual branch currents (bottom). Calculations appear on the following page.

	Column A	Column B	Column C
Loop 1	$(1+2)I_1$	$-2I_3$	$= -6$
Loop 2	$(3+4)I_2$	$-3I_3$	$= 6-7$
Loop 3	$(5+3+2)I_3$	$-3I_2-2I_1$	$= 7$

After collecting terms we will solve the equations of loops 1 and 2 for I_1 and I_2, respectively, and substitute these expressions into the equation for loop 3. This will leave I_3 as the only unknown in the equation for loop 3.

$$
\begin{array}{llll}
(1) & 3I_1 & -2I_3 & = -6 \\
(2) & 7I_2 & -3I_3 & = -1 \\
(3) & -2I_1 & -3I_2 & +10I_3 & = 7
\end{array}
$$

(1) $\quad I_1 = \dfrac{-6}{3} + \dfrac{2I_3}{3} = \dfrac{2}{3}I_3 - 2$

(2) $\quad I_2 = \dfrac{-1}{7} + \dfrac{3I_3}{7} = \dfrac{3}{7}I_3 - \dfrac{1}{7}$

(3) $\quad -2\left(\dfrac{2}{3}I_3 - 2\right) - 3\left(\dfrac{3}{7}I_3 - \dfrac{1}{7}\right) + 10I_3 = 7$

$$\left(10 - \dfrac{4}{3} - \dfrac{9}{7}\right)I_3 = 7 - 4 - \dfrac{3}{7}$$

$$I_3 = \dfrac{2.57}{7.38} = 0.348 \text{ A}$$

(1) $\quad 3I_1 = -6 + 2\,(0.348)$
$\qquad I_1 = -1.768 \text{ A}$

(2) $\quad 7I_2 = -1 + 3\,(0.348)$
$\qquad I_2 = 0.006 \text{ A}$

Nodal analysis is based on Kirchhoff's current law: *The algebraic sum of the currents leaving and entering any node of a circuit is zero.* A **node** is the junction of two or more branches. Computer circuit-analysis programs are often based on nodal analysis. Example circuit (Figure 3.12) appears on the next page. Here is the step-by-step procedure.

1. Convert all voltage sources to current sources.

2. Choose one node (usually ground) as the reference. Label all other nodes V_1, V_2, and so on.

3. Set up a table to form the node equations. There are three columns and a number of rows equal to the number of nodes (not including the reference node).

4. The column-A term is the sum of the conductances tied to node N times V_N.

5. The column-B terms are the conductances tied to node N and another node X, times V_X. Node X does not include the reference node. There may be several column-B terms. Each is subtracted from the column-A term.

6. The column-C term, to the right of the equals sign, is the algebraic sum of all current sources tied to node N. A source is termed positive if it supplies current to the node and negative if it takes current away.

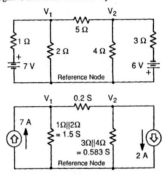

Figure 3.12   Node analysis: original circuit, above, and converted to conductance units, below.

	Column A		Column B		Column C
Node 1	$(1.5 + 0.2)V_1$	$-$	$0.2\,V_2$	$=$	7
Node 2	$(0.583 + 0.2)V_2$	$-$	$0.2\,V_1$	$=$	-2
(1)	$1.7\,V_1$	$-$	$0.2\,V_2$	$=$	7
(2)	$-0.2\,V_1$	$+$	$0.783\,V_2$	$=$	-2

Multiplying (2) by 8.5 and adding the result to (1):

$$-1.7\,V_1 \quad + \quad 6.656\,V_2 \quad = \quad -17$$
$$6.456\,V_2 \quad = \quad -10$$
$$V_2 \quad = \quad -1.549 \text{ V}$$

Substituting this V_2 into (1) yields $V_1 = 3.935$ V.

3.4 AC CIRCUIT ANALYSIS

This section is limited to circuits containing linear resistive and reactive elements and driven by sinusoidal waveforms of a single fixed frequency.

Circuit simplification. Elementary ac circuits that can be reduced to a single resistance in series or in parallel with a single reactance may be solved using only Ohm's law and the resistance-reactance combination formula. In Figure 3.13, below, we determine the output across a series-tuned circuit (a) shunted by a load resistance R_L, at a frequency 1% above resonance. Generator resistance r_g and coil-winding resistance r_w are included. The circuit is redrawn in (b) to show how the Thévenin equivalent at (c) is obtained. Calculations appear on the following page.

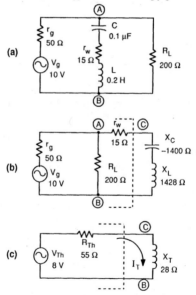

Figure 3.13 AC circuit simplification.

$$f_r = \frac{1}{2\pi\sqrt{LC}} = \frac{1}{2\pi\sqrt{0.2\times0.1\times10^{-6}}} = 1125 \text{ Hz}$$

Let $f = 1.01 f_r = 1137$ Hz.

$$X_L = 2\pi f L = 2\pi \times 1137 \times 0.2 = 1428 \ \Omega$$

$$X_C = \frac{-1}{2\pi f C} = \frac{-1}{2\pi \times 1137 \times 0.1 \times 10^{-6}} = -1400 \ \Omega$$

$$X_T = X_L + X_C = 1428 - 1400 = 28 \ \Omega$$

$$V_{Th} = V_G \frac{R_L}{R_L + r_g} = 10 \ \frac{200}{200 + 50} = 8.0 \text{ V}$$

$$R_{Th} = r_w + R_L \| r_g = 15 + 200\|50 = 55 \ \Omega$$

$$Z_T = \sqrt{R_{Th}^2 + X_T^2} = \sqrt{55^2 + 28^2} = 61.7 \ \Omega$$

$$I_T = I_{X(T)} = I_{CB} = I_{ACB} = \frac{V_{Th}}{Z_T} = \frac{8.0}{61.7} = 0.130 \text{ A}$$

$$Z_{ACB} = \sqrt{r_w^2 + X_T^2} = \sqrt{15^2 + 28^2} = 31.8 \ \Omega$$

$$V_{AB} = I_T Z_{ABC} = 0.130 \times 31.8 = 4.12 \text{ V}$$

Series-parallel R-X conversions (see page 23) often permit a more complex circuit to be reduced to single R and X components. The example of Figure 3.14 on the next page demonstrates that the output of a Pi-section L-C low-pass filter is down by a factor of 8 at twice the cutoff frequency. This represents three −6 dB cuts, or a −18 dB/octave roll-off. In the passband, r_g and R_L divide the 200 V, leaving $V_o' = 100$ V. At $2 f_c$ we expect $V_o = 1/8 \ V_o' = 12.5$ V. The circuit will be simplified in four stages. Voltage C-A is found in stage (d) and transferred to C-A in stage (b) where voltage division leads directly to V_{B-A}, which is V_o.

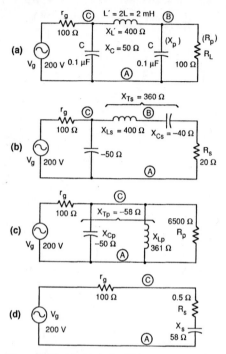

Figure 3.14 AC circuit simplification by series-parallel R-X conversions.

Stage (a):

$$f_c = \frac{1}{2\pi\sqrt{LC}} = \frac{1}{2\pi\sqrt{0.001 \times 0.1 \times 10^{-6}}} = 15\,915 \text{ Hz}$$

$$2f_c = 31\,830 \text{ Hz}$$

$$X_C = \frac{1}{2\pi f C} = \frac{1}{2\pi \times 31\,830 \times 0.1 \times 10^{-6}} = 50 \, \Omega$$

$$X_L' = 2\pi f L' = 2\pi \times 31\,830 \times 0.002 = 400 \, \Omega$$

Stage (b):

$$R_s = X_p \frac{X_p R_p}{X_p^2 + R_p^2} = 50 \frac{50 \times 100}{50^2 + 100^2} = 20 \ \Omega$$

$$X_{Cs} = -R_p \frac{X_p R_p}{X_p^2 + R_p^2} = -100 \frac{50 \times 100}{50^2 + 100^2} = -40 \ \Omega$$

$$X_{Ts} = X_{Ls} + X_{Cs} = 400 - 40 = 360 \ \Omega$$

Stage (c):

$$R_p = \frac{X_{Ts}^2 + R_s^2}{R_s} = \frac{360^2 + 20^2}{20} = 6500 \ \Omega$$

$$X_{Lp} = \frac{X_{Ts}^2 + R_s^2}{X_{Ts}} = \frac{360^2 + 20^2}{360} = 361 \ \Omega$$

$$X_{Tp} = \frac{X_{Lp} X_{Cp}}{X_{Lp} + X_{Cp}} = \frac{361(-50)}{361 - 50} = -58 \ \Omega$$

Stage (d):

$$R_s = X_{Tp} \frac{X_{Tp} R_p}{X_p^2 + R_p^2} = 58 \frac{58 \times 6500}{58^2 + 6500^2} = 0.5 \ \Omega$$

$$X_s = R_p \frac{X_{Tp} R_p}{X_{Tp}^2 + R_p^2} = 6500 \frac{58 \times 6500}{58^2 + 6500^2} = 58 \ \Omega$$

$$V_{CA} = V_g \frac{Z_{CA}}{Z_T} = 200 \frac{\sqrt{0.5^2 + 58^2}}{\sqrt{100.5^2 + 58^2}} = 100 \ V$$

Stage (b):

$$V_{BA} = V_{RL} = V_{CA} \frac{Z_{BA}}{Z_{CBA}} = 100 \frac{\sqrt{20^2 + 40^2}}{\sqrt{20^2 + 360^2}} = 12.4 \ V$$

AC analysis by complex algebra. The network theorems of Section 3.2 are valid for ac resistive-reactive circuits provided that the quantities R, V, and I are replaced by the complex quantities Z, V, and I. Section 1.7 gives the form and rules for manipulation of complex quantities. It bears repeating that complex addition $(V_1 + V_2)$ is done with the quantities in rectangular form. Simply summing the magnitudes will produce incorrect results unless the phase angles are the same. Multiplication and division $(I \times Z)$ are done in polar form. Multiplying and dividing magnitudes alone will produce results of correct magnitude, but phase-angle information, which may be needed for later additions, will be lost.

A **Wien-bridge notch filter** with a load resistance is analyzed at $f = 2/3\ f_{notch}$ in Figure 3.15, below. The circuit is reduced to a single-loop Thévenin equivalent with the aid of series-parallel R-X conversion (see page 23) and complex algebra. Every trick in the book should be used to simplify ac circuits before actually starting the solution, since complex algebra usually entails an immense amount of work. Let $R = 2\ k\Omega$ and $C = 0.1\ \mu F$. The notch frequency and reactance at the operating frequency are determined first:

$$f_c = \frac{1}{2\pi RC} = \frac{1}{2\pi \times 2000 \times 0.1 \times 10^{-6}} = 796\ \text{Hz}$$

$$f = 2/3 f_c = 531\ \text{Hz} \qquad\qquad X_C = 3000\ \Omega\ \text{at}\ f$$

The circuit is simplified in six stages, (a) through (f). The problem will be done with V in volts, R and X in kilohms, and I in milliamperes. Calculations are completed after the figures.

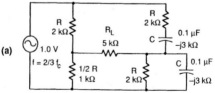

Figure 3.15 Wien-bridge analysis by complex algebra.

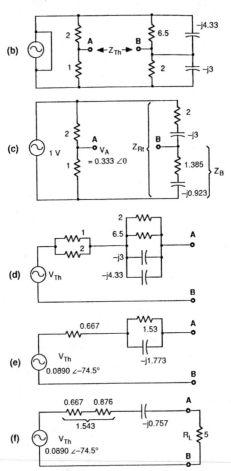

Figure 3.15 (continued)

107

V_B is found from Figure 3.15 (c), on the previous page:

$$Z_{Rt} = 3.385 - j\,3.923 = 5.182\angle{-49.2°}\ \Omega$$

$$I = \frac{V}{Z_{Rt}} = \frac{1\angle 0}{5.182\angle 49.2°} = 0.1930\angle 49.2°\ A$$

$$V_B = I Z_B = 0.1930\angle 49.2° \times 1.664\angle{-33.7°}$$

$$= 0.3212\angle 15.5° = (0.3085 + j\,0.0858)\ V$$

$$V_{Th} = V_A - V_B = (0.3333 + j\,0) - (0.3095 + j\,0.0858)$$

$$= 0.0238 - j\,0.0858 = 0.0890\angle{-74.5°}\ V$$

The current and load voltage are found from Figure 3.15 (f):

$$I = \frac{V_{Th}}{Z_{Rt}} = \frac{0.0238 - j\,0.0858}{6.543 - j\,0.757}$$

$$= \frac{0.0890\angle{-74.5°}}{6.587\angle{-6.6°}} = 0.0135\angle{-67.9°}\ A$$

$$V_{RL} = I R_L = 0.0135\angle{-67.9°} \times 5 = 0.0675\angle{-67.9°}\ V$$

The 5-kΩ load resistor thus drops V_{RL} to 76% of the no-load V_{Th} and shifts the phase of the output 6.6°.

Mesh and nodal analysis for ac circuits follow the rules given in Section 3.3 for dc circuits, where R, V, and I become the complex quantities Z, V, and I. The mathematics is quite tedious, and the chances of a human operator analyzing a nontrivial problem without error are slim. In a thoroughly debugged computer program, however, the method is outstanding.

As an example of ac mesh analysis, the twin-tee notch filter shown in Figure 3.16 on the following page will be analyzed at a frequency $f = 2/3\ f_c$. Notice that the circuit does not lend itself to reduction by series-parallel R-X conversion or Thévenin's theorem.

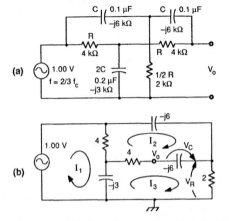

Figure 3.16 Twin-tee notch filter circuit (a), and as set up for loop analysis at (b).

Three equations are set up, based on Kirchhoff's voltage law, that the sum of the voltage drops around each of the three loops must equal the sum of any voltage rises (sources) in that loop. Voltage for each element is $I \times R$ or $I \times X$.

	Column A	Column B_1	Column B_2	C
Loop 1	$(4 - j3)I_1 -$	$(4)I_2 -$	$(-j3)I_3$	$= 1$
Loop 2	$(8 - j12)I_2 -$	$(4)I_1$	$(4 - j6)I_3$	$= 0$
Loop 3	$(6 - j9)I_3 -$	$(-j3)I_1 -$	$(4 - j6)I_2$	$= 0$

Eqn 1	$(4 - j3)I_1 +$	$(-4)I_2 +$	$(j3)I_3$	$= 1$
Eqn 2	$(-4)I_1 +$	$(8 - j12)I_2 +$	$(-4 + j6)I3$	$= 0$
Eqn 3	$(j3)I_1 +$	$(-4 + j6)I_2 +$	$(6 - j9)I_3$	$= 0$

These three equations will be solved by determinants, with D representing the common denominator, N_2 the numerator in the solution for I_2, and N_3 the numerator in the solution for I_3.

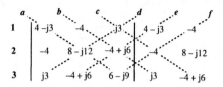

	a	b	c	d	e	f
1	$4-j3$	-4	$j3$	$4-j3$	-4	
2	-4	$8-j12$	$-4+j6$	-4	$8-j12$	
3	$j3$	$-4+j6$	$6-j9$	$j3$	$-4+j6$	

$$D = (a_1a_2a_3) + (b_1b_2b_3) + (c_1c_2c_3)$$
$$- (d_1d_2d_3) - (e_1e_2e_3) - (f_1f_2f_3)$$

The rectangular quantities are converted to polar form for multiplication, and then back to rectangular form:

$$D = (-672 - j396) + (72 + j48) + (72 + j48)$$
$$-(-72 + j108) - (-224 - j132) - (96 - j144)$$

$$= -328 - j132 = 353.6\angle{-158.1°} \ (k\Omega)^3$$

The objective will now be to solve for the currents in loop 3 and loop 2, so that V_o can be obtained by adding $V_R + V_C$, as labeled in Figure 3.16 on the preceding page. First solving for I_3:

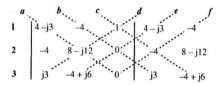

	a	b	c	d	e	f
1	$4-j3$	-4	1	$4-j3$	-4	
2	-4	$8-j12$	0	-4	$8-j12$	
3	$j3$	$-4+j6$	0	$j3$	$-4+j6$	

$$N_3 = 0 + 0 + (16 - j24) - (36 + j24) - 0 - 0 = -20 - j48$$
$$= -52.0\angle{-112.6°} \ (k\Omega)^2 \ V$$

$$I_3 = \frac{N_3}{D} = \frac{52.0\angle{-112.6°}}{353.6\angle{-158.1°}} = 0.147\angle{45.5°}$$

$$= (0.103 + j\,0.105) \ mA$$

Now solving for I_2, and proceeding to find V_o:

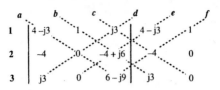

$$N_2 = 0 + (-18 - j12) + 0 - 0 - 0 - (-24 + j36)$$
$$= 6 - j48 = 48.4 \angle -82.9° \ (k\Omega)^2 \ V$$

$$I_2 = \frac{N_2}{D} = \frac{48.4 \angle -82.9°}{353.6 \angle -158.1°} = 0.137 \angle 75.2°$$
$$= (0.035 + j0.132) \ mA$$

$$V_R = I_3 R = (0.103 + j0.105) \ (2) = (0.206 + j0.210) \ V$$

$$V_C = (I_3 - I_2) \ X_C = [(0.103 + j0.105)$$
$$- (0.035 + j0.132)] \cdot [-j6] = (-0.162 - j0.408) \ V$$

$$V_o = V_R + V_C = (0.206 + j0.201) + (-0.162 - j0.408)$$
$$= 0.044 - j0.198 = 0.203 \angle -77.5° \ V$$

The results show the output at $2/3 \ f_c$ to be down by 13.9 dB and to be lagging the source by $77.5°$.

3.5 NONSINUSOIDAL EXCITATION

The square wave is the most common nonsinusoidal function, and many circuits can be reduced to single R-L or R-C branches driven by a square wave. These can be analyzed by elementary methods.

Short-time-constant R-L and R-C circuits, where the time constant is much less than the switching rate ($\tau \ll t$), can be analyzed with the formulas of Section 1.4.

Long-time-constant R-C circuits ($t \ll RC$) often can be analyzed with the following tools:

1. Thévenin's theorem.
2. $Q_{chg} = Q_{dischg}$ or $I_{chg} t_{chg} = I_{dischg} t_{dischg}$.
3. $C \Delta V = I \Delta t$, where ΔV is change in capacitor voltage, I is capacitor current (assumed to be reasonably constant), and Δt is time of charge or discharge ($\Delta t \ll \tau$).

Figure 3.17, below, gives an example in which an 11-V, 1-ms pulse with a 10-ms repetition time and 5-Ω internal resistance drives a half-wave rectifier and simple capacitor filter. We will find the V_o waveform. Note that $\tau \gg t$ for both charge and discharge.

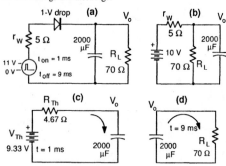

Figure 3.17 Original pulse rectifier (a), set up for Thévenin (b), charging (c), and discharging circuit (d).

$$I_{chg} = \frac{V_{Th} - v_o}{R_{Th}} \qquad\qquad I_{dischg} = \frac{v_o}{R_L}$$

$$I_{chg} t_{on} = I_{dischg} t_{off} \qquad \frac{V_{Th} - v_o}{R_{Th}} t_{on} = \frac{v_o}{R_L} t_{off}$$

$$v_o = \frac{V_{Th}}{1 + \frac{R_{Th} t_{off}}{R_L t_{on}}} = \frac{9.33}{1 + \frac{4.67 \times 9 \text{ ms}}{70 \times 1 \text{ ms}}} = 5.83 \text{ V dc avg}$$

$$\Delta v_o = \frac{I_L t_{off}}{C} = \frac{v_o t_{off}}{R_L C} = \frac{5.83 \times 9 \text{ ms}}{70 \times 2 \text{ mF}} = 0.37 \text{ V p-p}$$

Long-time-constant R-L circuits ($t \gg L/R$) in which the inductor does not saturate, can be analyzed with the following formulas:

1. Thévenin's theorem.

2. $V_{chg} t_{chg} = V_{dischg} t_{dischg}$; which states that the average voltage across an inductor must be zero (positive half-cycle equals negative half-cycle).

3. $L \, \Delta I = V \, \Delta t$, where ΔI is the change in current through the inductor in time Δt, and V is the voltage across the inductor during Δt. Voltage V and bulk current I are assumed constant over Δt since $\tau \gg t$.

Figure 3.18, below, shows a pulse transformer switched on for 0.3 ms by a transistor. Load voltage, inductor current, and supply current are found.

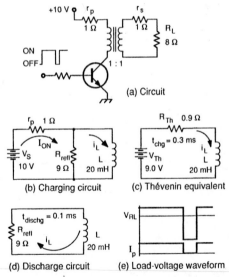

(a) Circuit

(b) Charging circuit

(c) Thévenin equivalent

(d) Discharge circuit

(e) Load-voltage waveform

Figure 3.18 Transistor-driven pulse transformer.

From Figure 3.18 (d):

$$I_{L(avg)} = \frac{V_{dischg}}{R_{refl}} ; \text{ but } V_{dischg} = \frac{V_{chg} t_{chg}}{t_{dischg}} ; \text{ and}$$

$$V_{chg} = V_{Th} - V_{R\,Th} = V_{Th} - I_{L(avg)} R_{Th}, \text{ so}$$

$$I_{L(avg)} = \frac{\left(V_{Th} - I_{L(avg)} R_{Th}\right) t_{chg}}{R_{refl} t_{dischg}} ; \text{ which becomes}$$

$$I_{L(avg)} = \frac{V_{Th} t_{chg}}{R_{refl} t_{dischg} + R_{Th} t_{chg}} ;$$

$$= \frac{9 \times 0.3}{(9 \times 0.1) + (0.9 \times 0.3)} = 2.31 \text{ A}$$

$$V_{L(chg)} = V_{Th} - I_{L(avg)} R_{Th} = 9 - (2.31 \times 0.9) = 6.92 \text{ V}$$

$$V_{L(dischg)} = I_{L(avg)} R_{refl} = 2.31 \times 9 = 20.8 \text{ V}$$

$$V_{RL(chg)} = V_{L(chg)} \frac{R_L}{r_s + R_L} = 6.92 \times \frac{8}{9} = 6.15 \text{ V}$$

$$V_{RL(dischg)} = V_{L(dischg)} \frac{R_L}{r_s + R_L} = 20.8 \times \frac{8}{9} = 18.5 \text{ V}$$

$$V_{R(refl)} = V_s - V_{Rp} \quad \text{and}$$

$$(I_{ON} - I_L) R_{refl} = V_s - I_{ON} R_p \quad \text{so}$$

$$I_{ON} = \frac{V_s + I_L R_{refl}}{R_{refl} + R_p} = \frac{10 + (2.31 \times 9)}{9 + 1} = 3.08 \text{ A}$$

$$\Delta I = \frac{V \, \Delta t}{L} = \frac{6.92 \times 0.3 \text{ ms}}{0.02} = 0.10 \text{ A}$$

$$P_T = I_{ON} V_s \eta_{chg} = 3.08 \times 10 \times \frac{3}{4} = 23.1 \text{ W}$$

$$P_{RL} = \frac{V_{RL}^2}{R_L} \eta_{chg} + \frac{V_{RL}^2}{R_L} \eta_{dischg}$$

$$= \frac{6.15^2}{8} \times \frac{3}{4} + \frac{18.5^2}{8} \times \frac{1}{4} = 14.2 \text{ W}$$

3.6 REAL-TRANSFORMER EQUIVALENT CIRCUITS

Power transformers and audio transformers behave according to the simple circuit model of Figure 3.19. Except for very large transformers, skin effect is negligible, so measured dc winding resistance may be used for r_p and r_s.

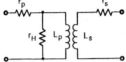

Figure 3.19 Equivalent circuit for close-coupled low-frequency transformer.

Inductances L_p and L_s behave like the ideal transformer of Section 1.9. Any voltage of whatever waveshape appearing across L_p will also appear, scaled by a factor of n, across L_s. Any current through L_s will appear, scaled by a factor of n, in L_p. Impedances across L_s are reflected across L_p by a factor of $1/n^2$. The load across L_s appears to be driven by the impedance that drives L_p, scaled by a factor of n^2.

Resistance r_H represents hysteresis loss in the core. It may be nonlinear, but is usually many times higher than X_{Lp} and causes less power loss than r_p or r_s.

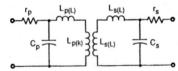

Figure 3.20 Equivalent circuit for high-frequency and loose-coupled transformers.

Video and rf transformers behave according to the more complex circuit of Figure 3.20, above. Skin effect (page 17) will increase r_p and r_s significantly over their dc values. The coils are assumed to have an air or slug-type ferrite core, so they are not perfectly coupled magnetically. The subscripts k and L represent coupled and leakage inductances. C_p and C_s represent stray winding capacitances, which range from 0.01 to 0.2 pF per turn.

$$L_{p(L)} + L_{p(k)} = L_p \qquad\qquad L_{s(L)} + L_{s(k)} = L_s$$

$$k = \frac{L_{p(k)}}{L_p} = \frac{L_{s(k)}}{L_s}$$

Typical coefficients of coupling k for air-core coils

Bifilar (two strands wound together)	$k \approx 0.95$
Single layer, primary over secondary, 20 turns	$k \approx 0.90$
End-to-end coils; length = 1/2 diameter	$k \approx 0.35$
End-to-end coils; length = 2 diameters	$k \approx 0.10$
In-line; separated by one coil length	$k \approx 0.02$

3.7 POWER-SUPPLY CIRCUIT ANALYSIS

Common rectifier circuits and their output voltages are given in Figure 3.21 on the following two pages. Ripple frequencies and typical ratios of rms secondary current to dc load current are also given. The circuits at (a), (b), and (c) can feed choke-input filters instead of the capacitors shown, in which case I_s will be about 70% of the values listed. The diode peak-inverse voltages listed are values encountered in normal service and should be multiplied by a safety factor of two, minimum, to cover for line transients. Capacitors should be selected with working voltages 1.5 times the operating voltages shown.

Rectifier output voltage will drop under load because of the IR drop when the first capacitor is charged by pulses of current through the transformer winding resistance. An estimate of this voltage drop can be given as

$$V_{drop} \approx 2.5\, I_{s(rms)} r_w$$

where $I_{s(rms)}$ is obtained from I_L and the various equations given with the figures. The factor of 2.5 indicates that $I_{s(pk)}$ is typically $2.5 \times I_{s(rms)}$. Secondary *plus* reflected primary resistance is included in r_w:

$$r_w = R_s + R_p \left(\frac{V_s}{V_p}\right)^2$$

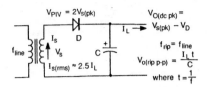

(a) Half-wave rectifier

$V_{PIV} = 2V_{s(pk)}$

$V_{O(dc\ pk)} = V_{s(pk)} - V_D$

$f_{rip} = f_{line}$

$V_{o(rip\ p-p)} = \dfrac{I_L t}{C}$

where $t = \dfrac{1}{f}$

$I_{s(rms)} \approx 2.5\,I_L$

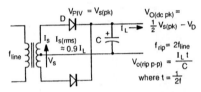

(b) Full-wave center-tapped rectifier.

$V_{PIV} = V_{s(pk)}$

$V_{O(dc\ pk)} = \dfrac{1}{2}V_{s(pk)} - V_D$

$f_{rip} = 2f_{line}$

$V_{o(rip\ p-p)} = \dfrac{I_L t}{C}$

where $t = \dfrac{1}{2f}$

$I_{s(rms)} \approx 0.9\,I_L$

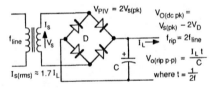

(c) Full-wave bridge rectifier.

$V_{PIV} = 2V_{s(pk)}$

$V_{O(dc\ pk)} = V_{s(pk)} - 2V_D$

$f_{rip} = 2f_{line}$

$V_{o(rip\ p-p)} = \dfrac{I_L t}{C}$

where $t = \dfrac{1}{2f}$

$I_{s(rms)} \approx 1.7\,I_L$

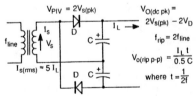

(d) Full-wave voltage-doubler circuit.

$V_{PIV} = 2V_{s(pk)}$

$V_{O(dc\ pk)} = 2V_{s(pk)} - 2V_D$

$f_{rip} = 2f_{line}$

$V_{o(rip\ p-p)} = \dfrac{I_L t}{0.5\,C}$

where $t = \dfrac{1}{2f}$

$I_{s(rms)} \approx 5\,I_L$

Figure 3.21 Rectifier circuit summary.

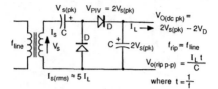

(e) Half-wave voltage-doubler circuit.

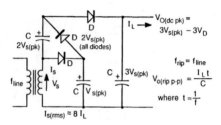

(f) Half-wave voltage tripler.

Figure 3.21 (Continued).

Ripple filters are summarized in the figures on the following page. The first capacitor filter C_1 is viewed as a charge reservoir governed by the equation

$$C_1 \Delta V_o = I_L \Delta t$$

where Δt is the time between charging pulses from the rectifier. The choke or filter resistance and the following capacitor C_2 are viewed as a voltage divider. X_L is assumed to be much greater than X_C, which is true in all practical cases. The factor 0.8 which appears with several of the figures stems from the fact that the ac waveform at the input of this voltage divider is a sawtooth, and the fundamental-frequency sine-wave peak is about 0.8 V_{pk}. The choke is often replaced by a resistor, which is much smaller, lighter, and less expensive.

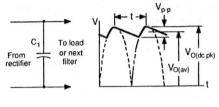

$$V_{\text{p-p}} = \frac{I_L t}{C_1} = \frac{I_L}{f_{\text{rip}} C_1} = \frac{V_{O(\text{av})}}{f_{\text{rip}} R_L C_1}$$

$$V_{O(\text{av})} = V_{O(\text{dc pk})} - \frac{1}{2} V_{\text{p-p}}$$

Figure 3.22 Single-capacitor filter following rectifier: waveforms and formulas.

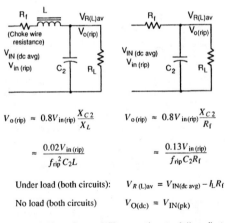

$$V_{o(\text{rip})} \approx 0.8 V_{\text{in(rip)}} \frac{X_{C2}}{X_L} \qquad V_{o(\text{rip})} \approx 0.8 V_{\text{in(rip)}} \frac{X_{C2}}{R_f}$$

$$\approx \frac{0.02 V_{\text{in(rip)}}}{f_{\text{rip}}^2 C_2 L} \qquad \approx \frac{0.13 V_{\text{in(rip)}}}{f_{\text{rip}} C_2 R_f}$$

Under load (both circuits): $\quad V_{R(L)\text{av}} = V_{\text{IN(dc avg)}} - I_L R_f$

No load (both circuits) $\quad V_{O(\text{dc})} = V_{\text{IN(pk)}}$

Figure 3.23 Second filter section to follow first capacitor filter: circuits and formulas.

Zener regulators provide load and line regulation as well as ripple reduction. Figure 3.24, below, gives the basic circuit. The regulation formulas neglect temperature effects on V_Z and are therefore valid only when self-heating is negligible. Unfortunately, temperature effects generally predominate over zener-resistance effects unless the power dissipated in the diode is less than 1/10 of its rated power.

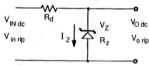

Figure 3.24   Elementary zener voltage regulator.

$$R_d = \frac{V_{IN\,dc} - V_Z}{I_Z + I_L} \qquad V_{o\,rip} = \frac{V_{in\,rip} - R_z}{R_d + R_z}$$

$$\eta_{LOAD\,REG} = \frac{R_z}{R_L} \qquad \eta_{LINE\,REG} = \frac{V_{IN\,dc}\,R_z}{V_Z\,R_d}$$

3.8   AMPLIFIER CIRCUIT ANALYSIS

Formulas for the analysis of transistor amplifiers can be developed from circuit-analysis techniques coupled with the following facts:

• Base current is many times smaller than emitter current (typically 40 to 300 times smaller for small-signal silicon transistors). Base current can thus be neglected and base bias determined by a voltage divider R_{B1}-R_{B2} if these resistors are low enough in value to carry a current several times larger than the base current.

• The forward-biased base-emitter junction presents a dynamic (ac) resistance r_e given approximately by

$$r_e = \frac{30\,mV}{I_E}$$

The theoretical value determined by Shockley is 26 mV, but in practice the equation is subject to errors on the order of ±40%, so the rounding to 30 mV is justified.

• Resistance looking into the base is β times the resistance in the emitter line, including r_e.

- Resistance looking into the emitter is r_e, plus the resistance looking out of the base divided by β.

- Collector current is nearly equal to emitter current for both ac and dc. This allows voltage-gain formulas to be developed in a manner similar to voltage-divider formulas:

$$i_c = \frac{v_c}{r_c}, \quad i_e = \frac{v_e}{r_e}, \quad \text{and} \quad i_c \approx i_e \quad \text{so} \quad \frac{v_c}{v_e} = \frac{r_c}{r_e}$$

where r_c and r_e are resistances through which the collector and emitter signal currents flow.

- Resistance seen looking into the collector is very high since it is a reverse-biased diode.

- Output peak signal swing cannot be greater than the swing from the rest (quiescent) point to saturation turn-on, at which $V_{CE} \to 0$ V.

- Output peak signal cannot be greater than what the bias rest current can develop across the output load when the transistor swings to cutoff ($I_C \to 0$ A).

The stabilized common-emitter amplifier shown in Figure 3.25 on the following page provides an inverted output signal typically 5 to 100 times the voltage of the ac input signal. Input and output impedance levels are comparable. Input impedance is somewhat affected by transistor beta, especially in high-gain circuits, but other operating characteristics are essentially beta-independent, provided that at least one volt is allowed across the emitter resistor, and that beta is relatively high. Typically $\beta_{min} \geq 40$ is adequate.

DC bias formulas:

1. $V_B = \dfrac{V_{CC} R_{B2}}{R_{B1} + R_{B2}}$

    provided that $\beta (R_{E1} + R_{E2}) \gg R_{B1} \| R_{B2}$.

2. $V_E = V_B - V_{BE}$

    where $V_{BE} \approx 0.6$ V for silicon, 0.2 V for germanium.

3. $I_C \approx I_E = \dfrac{V_E}{R_E}$

121

AC signal formulas:

1. $r_e \approx \dfrac{30 \text{ mV}}{I_E}$ [subject to ±40% errors]

2. $Z_{in} = R_{B1} \parallel R_{B2} \parallel \beta(r_e + R_{E1})$

3. $A_v = \dfrac{V_o}{V_{in}} = \dfrac{R_C \parallel R_L}{r_e + R_{E1}}$

4. $Z_o = R_C$

5. $V_{o(max)\text{p-p}} \leq 2V_{CE}$
    [negative peak limited by transistor saturation]

    $V_{o(max)\text{p-p}} \leq 2I_C(R_C \parallel R_L)$
    [positive peak limited by bias-current cutoff]

6. $f_{low} \geq \dfrac{1}{2\pi C_{in}(Z_s + Z_{in})}$ −3 dB frequency
determined by

    $f_{low} \geq \dfrac{1}{2\pi C_E(r_e + R_{E1})}$ highest of three
calculations

    $f_{low} \geq \dfrac{1}{2\pi C_o(R_C + R_L)}$

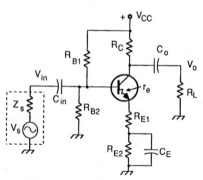

Figure 3.25 Stabilized common-emitter amplifier.

122

Collector self-bias, shown in Figure 3.26, below, is sometimes used to improve bias stability (to achieve a constant V_C in the face of β and temperature variations). The Miller effect lowers input impedance considerably, however, because the output signal appears across R_{B1}, forcing the source to deliver more current to it. Other properties are the same as for the stabilized common-emitter amplifier of the previous page. Note that $R_{E1} = 0$ for maximum gain (but even lower Z_{in}) in the example circuit shown. The new bias and input-impedance formulas are:

$$V_C = \frac{R_E V_{CC} + R_C V_{BE}}{\dfrac{R_{B2} R_C}{R_{B1} + R_{B2}} + R_E}$$

$$Z_{in} = R_{B2} \| \frac{R_{B1}}{A_v} \| \beta(R_{E1} + r_e)$$

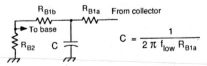

Figure 3.26 Collector self-bias for stability.

The input impedance of a collector-self-biased amplifier can be raised to nearly its full former value by filtering the ac signal from the collector voltage with a capacitor before feeding the dc back to the base, as shown in the partial schematic of Figure 3.27, below.

$$C = \frac{1}{2\pi f_{low} R_{B1a}}$$

Figure 3.27 *C* raises input impedance.

Bootstrapping, shown in Figure 3.28, below, is used to increase input impedance. It is effective only if $R_{E1} \gg r_e$, dictating much less than maximum voltage gain. Bias stability suffers, since base bias current is obtained from $(R_{B1} \parallel R_{B2}) + R_{B3}$, and R_{B3} is relatively large.

$$Z_{in} = R_{B3} \frac{r_e + R_{E1}}{r_e} \parallel \beta(R_{E1} + r_e)$$

provided that $R_{B3} \gg R_{B1} \parallel R_{B2}$, and

$$R_{B1} \parallel R_{B2} \gg R_{E1}$$

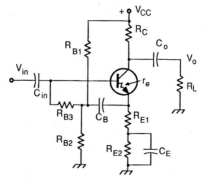

Figure 3.28 Bootstrapping for high input impedance.

The emitter follower, shown in Figure 3.29 on the next page, presents a high input impedance to the source and a low Z_o to the load. Bootstrapping components R_{B3} and C_B are often used to eliminate R_{B1} and R_{B2} from the input-impedance equation, as in the figure above. Voltage gain is noninverting and slightly less than 1.

1. $V_E = \dfrac{V_{CC} R_{B2}}{R_{B1} + R_{B2}}$ 2. $I_E = \dfrac{V_E}{R_E}$

3. $A_v = \dfrac{R_E \parallel R_L}{r_e + R_E \parallel R_L}$ 4. $r_e \approx \dfrac{30 \text{ mV}}{I_E}$

5. $\quad Z_{in} = R_{B1} \| R_{B2} \| \beta(r_e + R_E \| R_L)$

6. $\quad Z_o = R_E \| \left(r_e + \dfrac{R_{B1} \| R_{B2} \| Z_s}{\beta} \right)$

7. $\quad f_{low} \geq \dfrac{1}{2 \pi C_{in}(Z_s + Z_{in})}$

8. $\quad f_{low} \geq \dfrac{1}{2 \pi C_o(Z_o + R_L)}$

9. $\quad V_{o(max)p\text{-}p} \leq 2 I_E(R_E \| R_L)$

10. $\quad V_{o(max)p\text{-}p} \leq 2 V_{CE}$

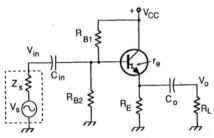

Figure 3.29 The emitter-follower circuit.

A transformer-coupled amplifier is shown in Figure 3.30 (a) on the following page. Figure 3.30 (b) gives an ac equivalent circuit with all resistances reflected to the collector. Figure 3.30 (c) shows the load lines for class-A and class-B operation. The transformer is assumed to have negligible leakage inductance ($k \approx 1$) but non-negligible winding resistance. This is true for most af transformers. Notice that a lower turns ratio n really does produce *higher* A_v, because r_{refl} in the collector line is raised by $1/n^2$. The I_Q formula is for optimum bias point where positive and negative signal swings are equal. Lower I_Q reduces $V_{o(max)}$. Higher I_Q simply wastes power.

1. $\quad A_v = \dfrac{V_o}{V_{in}} = \dfrac{R_L}{n(r_c + R_E)}$

2. $\quad I_{Q(\text{opt-class A})} = \dfrac{V_{CC} - V_{CE(\text{sat})}}{\dfrac{r_s + R_{L_s}}{n^2} + 2(R_E + r_p)}$

3. $\quad V_{o(\text{max p-p) class A}} = \dfrac{R_L(V_{CC} - V_{CE(\text{sat})})}{n(R_E + r_p) + \dfrac{r_s + R_L}{2n}}$

4. $\quad V_{o(\text{max p-p) class B}} = \dfrac{R_L(V_{CC} - V_{CE(\text{sat})})}{n(R_E + r_p) + \dfrac{r_s + R_L}{n}}$

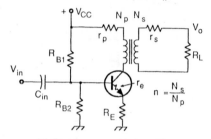

(a) Transformer-coupled amplifier.

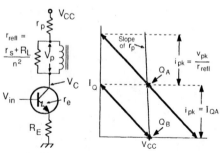

(b) Equivalent circuit. (c) Load lines.

Figure 3.30. Transformer-coupled amplifier analysis.

FET amplifiers may be analyzed by formulas similar to those used for bipolar amplifiers. Here are the main differences:

- Gate-to-source voltage (instead of base current) controls device current. V_{GS} may be from –0.5 to –5 V for typical low-power N-channel junction FETs (compared to a relatively constant 0.6 V for silicon NPN transistors).

- Gate input impedance is nearly infinite, compared to the bipolar transistor's input resistance of $\beta (r_e + R_{B1})$.

- Base-emitter junction resistance r_e is replaced by a fictitious source resistance $r_s = 1/y_{fs}$. Forward transfer admittance y_{fs} is sometimes called mutual conductance g_m or g_{fs}. It is the ac gain i_{out}/v_{in} of the FET, and is typically 3 to 15 mS for small-signal FETs.

- Minimum drain-to-source voltage $V_{DS(sat)}$ is typically 1 to 3 V at medium currents, compared to 0.1 to 0.3 V for $V_{CE(sat)}$ in bipolar transistors.

A common-source FET amplifier is shown in Figure 3.31 on the next page. Single-supply bias is possible, but it requires closely controlled FET parameters, a relatively high V_{DD} supply voltage, and a gate voltage divider R_{G1}-R_{G2}. Dual supplies give better bias stability and are to be preferred. C_{in} and R_G are shown to block dc from the input, but if the source has zero dc voltage and a dc path to ground, they can be omitted. Use absolute (unsigned) values in the following equations.

$$I_D = \frac{V_{SS} + V_{GS}}{R_S} \qquad V_D = V_{DD} - I_D R_D$$

$$Z_{in} = R_G \qquad A_v = \frac{R_D \| R_L}{r_s} = y_{fs}(R_D \| R_L)$$

$$V_{o(max)p-p} \le I_D(R_D \| R_L)$$

$$\le V_{DS} - V_{DS(sat)}$$

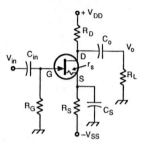

Figure 3.31 Basic common-source FET amplifier.

A source-follower FET amplifier is shown in Figure 3.32, below, with its relevant equations. The source follower is easier to bias and requires less supply voltage than the common-source amplifier. Since the gain of a common-source amplifier is not especially high anyway, designers often use a source follower as an input stage to obtain high input impedance, and follow it with bipolar common-emitter stages to obtain the desired voltage gain.

$$I_D = I_S = \frac{V_{SS} + V_{GS}}{R_S} \qquad V_S = V_{GS}$$

$$Z_{in} = R_G \qquad A_v = \frac{R_S \| R_L}{r_s + R_S \| R_L}$$

$$V_{o(max)p\text{-}p} \leq I_D(R_S \| R_L)$$
$$\leq V_{DS} - V_{DS(sat)}$$

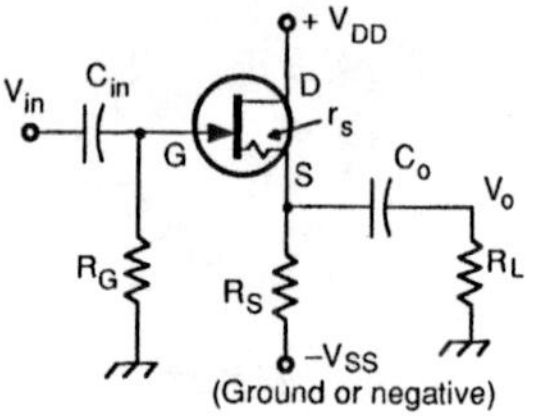

Figure 3.32 Basic source-follower amplifier.

128

Negative feedback is used to reduce amplifier distortion and increase amplifier bandwidth. It always decreases amplifier gain. It increases input impedance if the fed-back signal is applied in series with the input signal, but it decreases input impedance if the fed-back signal is applied in shunt with the input signal.

If the fed-back signal is obtained from the output voltage (by a voltage divider or step-down transformer), the amplifier's output impedance is lowered by negative feedback. This tends to make the output voltage a better reproduction of the input voltage. If the fed-back signal is obtained from the output current (in series with the load), output impedance is raised by negative feedback. This tends to make the load *current* a more faithful reproduction of the input voltage. In each case, the change in gain or impedance is by a factor of $(1 + A_{vo}B)$. This might be termed the "throw-away" factor, since the basic trade-off in negative feedback is to throw away gain to buy improvements in fidelity, impedance, bandwidth, or stability.

Figures 3.33 through 3.36 (following pages) and the accompanying equations summarize the four possible configurations of negative feedback and give example data to illustrate the effects. Characteristics of the amplifier proper, without the feedback circuit, are denoted by the subscripts "o" (open loop), "a" (inputs), and "b" (outputs). Closed-loop parameters, referring to characteristics of the complete circuit with feedback components connected, are denoted by the subscript "c." The fraction of output signal fed back is given as B (some texts use β.) Note that the term $A_{vo}B$ is always positive, since a negative A_{vo} requires a negative B to keep the feedback negative.

The input impedances affected by negative feedback are only those within the feedback loop. Impedances directly in series or directly shunted across the signal source appear as their normal values.

In the simple case where only one capacitance or inductance limits f_{high} or f_{low}, negative feedback extends bandwidth by the factor $(1 + A_{vo}B)$. Where two reactive elements cut off at the same frequency, the benefit factor is $\sqrt{1 + A_{vo}B}$. For three reactive elements cutting off at the same frequency the factor is $\sqrt[3]{1 + A_{vo}B}$.

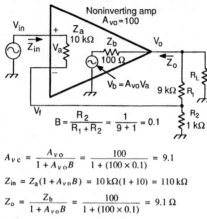

$$A_{vc} = \frac{A_{vo}}{1 + A_{vo}B} = \frac{100}{1 + (100 \times 0.1)} = 9.1$$

$$Z_{in} = Z_a(1 + A_{vo}B) = 10\ k\Omega(1 + 10) = 110\ k\Omega$$

$$Z_o = \frac{Z_b}{1 + A_{vo}B} = \frac{100}{1 + (100 \times 0.1)} = 9.1\ \Omega$$

Figure 3.33 Series-applied, voltage-derived feedback.

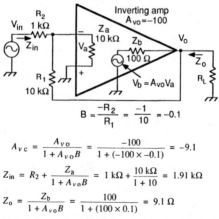

$$A_{vc} = \frac{A_{vo}}{1 + A_{vo}B} = \frac{-100}{1 + (-100 \times -0.1)} = -9.1$$

$$Z_{in} = R_2 + \frac{Z_a}{1 + A_{vo}B} = 1\ k\Omega + \frac{10\ k\Omega}{1 + 10} = 1.91\ k\Omega$$

$$Z_o = \frac{Z_b}{1 + A_{vo}B} = \frac{100}{1 + (100 \times 0.1)} = 9.1\ \Omega$$

Figure 3.34 Shunt-applied, voltage-derived feedback.

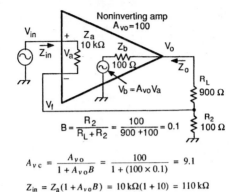

$$A_{vc} = \frac{A_{vo}}{1 + A_{vo}B} = \frac{100}{1 + (100 \times 0.1)} = 9.1$$

$$Z_{in} = Z_a(1 + A_{vo}B) = 10\text{ k}\Omega(1 + 10) = 110\text{ k}\Omega$$

$$Z_o = Z_b(1 + A_{vo}B) = 1000(1 + 10) = 11\text{ k}\Omega$$

Figure 3.35 Series-applied, current-derived feedback.

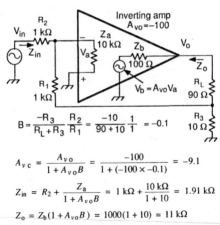

$$A_{vc} = \frac{A_{vo}}{1 + A_{vo}B} = \frac{-100}{1 + (-100 \times -0.1)} = -9.1$$

$$Z_{in} = R_2 + \frac{Z_a}{1 + A_{vo}B} = 1\text{ k}\Omega + \frac{10\text{ k}\Omega}{1 + 10} = 1.91\text{ k}\Omega$$

$$Z_o = Z_b(1 + A_{vo}B) = 1000(1 + 10) = 11\text{ k}\Omega$$

Figure 3.36 Shunt-applied, current-derived feedback.

Operational amplifier circuits are shown in Figures 3.37 (inverting) and 3.38 (noninverting). The inverting form is more popular because it can accept multiple inputs, and because the IC input remains at virtual ground level, minimizing problems from input stray capacitance. Both circuits respond to dc as well as to ac inputs.

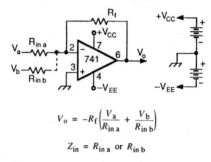

$$V_o = -R_f\left(\frac{V_a}{R_{in\,a}} + \frac{V_b}{R_{in\,b}}\right)$$

$$Z_{in} = R_{in\,a} \text{ or } R_{in\,b}$$

Figure 3.37 The inverting op-amp circuit is most popular. Note connection of two power supplies.

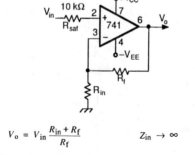

$$V_o = V_{in}\frac{R_{in} + R_f}{R_f} \qquad\qquad Z_{in} \to \infty$$

Figure 3.38 The noninverting op-amp circuit has very high Z_{in}. R_{saf} protects IC from input overvoltage.

3.9 TRIGGER AND POWER CIRCUITS

The **555 timer IC** appears in its two most popular applications in Figures 3.39 and 3.40, below.

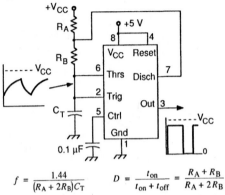

$$f = \frac{1.44}{(R_A + 2R_B)C_T} \qquad D = \frac{t_{on}}{t_{on} + t_{off}} = \frac{R_A + R_B}{R_A + 2R_B}$$

Figure 3.39 A 555 Oscillator circuit, with formulas for frequency and duty cycle (percent *on* time.)

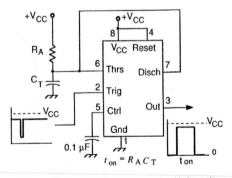

$$t_{on} \approx R_A C_T$$

Figure 3.40 A 555 one-shot circuit with formula.

133

Two common power circuits appear below. Figure 3.41 shows a variable dc supply using the popular 723 regulator IC. Figure 3.42 shows a standard lamp-dimmer using a triac and diac.

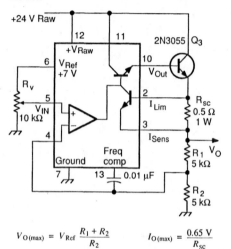

$$V_{O(max)} = V_{Ref} \frac{R_1 + R_2}{R_2} \qquad I_{O(max)} = \frac{0.65 \text{ V}}{R_{sc}}$$

Figure 3.41 A regulator circuit using the popular 723 IC. For low currents, omit Q_3 and connect pin 10 to 2.

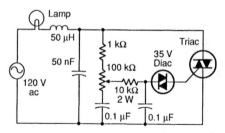

Figure 3.42 Standard lamp-dimmer circuit.

134

4

Units, Conversions, and Constants

4.1 THE INTERNATIONAL SYSTEM OF UNITS (SI)

Quantities are physical characteristics capable of being expressed numerically. Examples are length, pressure, and electric current. *Units* are arbitrary amounts that form the basis for measurement of quantities. Examples are centimeters, pounds per square inch, and amperes. The *Systeme Internationale d'Unites* (SI) has been developed to provide a single, well-defined, and universally accepted unit for each quantity.

Quantity symbols:

1. Quantity symbols consist of a single letter of the English or Greek alphabet, modified by subscripts and/or superscripts as appropriate.

2. When printed, quantity symbols (and mathematical variables) appear in *italic* (slanted) type. Subscripts that are quantity symbols in their own right are also italic. Other subscripts are roman (upright). Examples are

$$V_R, \ V_{\max}, \ I_x, \ I_{\text{CEO}}$$

3. Boldface italic type may be used to distinguish a vector quantity (I) from a scalar quantity (I). If lightface is used for vectors, then magnitudes should be distinguished by the *absolute* sign: $|I|$.

4. Because of the limited number of characters available, two quantities may be assigned the same letter symbol. To avoid confusion, an alternative letter symbol may be employed if one is listed, or the quantities may be differentiated by subscripts, or upper- and lowercase letters may be defined differently by the writer. In all cases the same quantity symbol should be retained throughout the work. Examples of such differentiations follow:

t (time), θ (temperature)

t (time), t_p (temperature)

t (time), T (temperature)

5. Several subscripts may be attached to a single quantity symbol, separated by a comma, hyphen, or parentheses if necessary for clarity. Multiple-level subscripts (subscripts attached to subscripts) are discouraged. A symbol with a superscript should be enclosed in parentheses before an exponent is added.

Electrical quantity symbols follow several additional conventions. Figure 4.1 illustrates some of them.

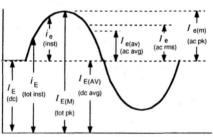

Figure 4.1 Quantity-symbol and subscript usage.

6. Uppercase (capital) letters are used for the quantity symbols for voltage, current, and power to designate dc, rms, average (av), maximum (m), or minimum (n, for *nadir*: lowest point) values. Uppercase subscripts to uppercase quantity symbols indicate dc values, or, with the subscripts M, AV, or N, respectively, the maximum, average, or minimum value of the total waveform. Lowercase subscripts to uppercase quantity symbols indicate the rms value of the ac wave, or with the subscript "m" the maximum value of the ac component of the total wave.

7. Lowercase (small) letters are used to designate the instantaneous value of a time-varying quantity. Here uppercase subscripts indicate the instantaneous value of the total waveform (dc + ac), and lowercase subscripts indicate the instantaneous value of the ac component only.

8. Double-uppercase subscripts designate the dc supply for the element indicated. Example: V_{CC} for the dc collector supply voltage.

9. Two-letter subscripts to the voltage symbol designate voltage from the first point to the second point as a reference. Single-letter subscripts to the voltage symbol may indicate voltage across the indicated device, or from the point designated to circuit ground as a reference.

Examples: V_{CE} (collector to emitter)
           V_Z (across the zener diode)
           V_C (collector to ground)
           V_o (output to ground, ac)

10. Hyphenated subscripts may be used where two elements have the same name:

Examples: $V_{B1\text{-}B2}$ (base 1 to base 2 of UJT)
           $V_{1B\text{-}2B}$ (base of the first transistor
                            to base of the second)

11. Conventional current (positive to negative through the load) is regarded as flowing *into* the terminal indicated by a subscript to the current symbol. Conventional current out of the terminal gives the quantity a negative sign.

    Figure 4.2, on the following page, shows this standard applied to one TTL gate driving another. The input delivers current back to the output of the driver in TTL, but the current leaving the input is designated as negative, and the same current entering the output is designated as positive.

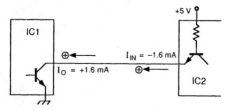

Figure 4.2. Current entering a device is positive. Current leaving a device is negative.

12. Bulk resistance is total voltage divided by total current, and is designated by uppercase R. Dynamic resistance ($\Delta V / \Delta I$) is designated by lowercase r. Lowercase r may also be used to designate inherent resistance of devices such as signal sources, inductors, and semiconductors.

Unit symbols consist of a letter or group of letters from the English and Greek alphabets, plus a few special symbols.

1. Unit symbols are printed in roman (upright) type. They are never given subscripts or superscripts.

2. Lowercase letters are used for unit symbols except where the symbol was derived from a proper name, in which case the first letter is capitalized.

3. A space is left between the number and the unit symbol.

4. Compound units, formed by multiplication and/or division of basic units, are common. Use a raised period to separate multiplications and a solidus (slash) or negative exponent to indicate division.

    Examples: $N \cdot m$ for newton-meters, and W/m^2
                    or $W \cdot m^{-2}$ for watts per square meter.

5. Clarity of meaning is often served by expressing certain quantities in "phantom" units. These should be reduced to strict SI units for purposes of calculation. Examples follow.

- Radian/second, cycle/second, and revolution/second reduce to $1/s$.

- Ampere·turn for magnetomotive force reduces to A.

- V/V, mV/V, and so on for gain, regulation, common-mode rejection ratio, and so on, reduce to a unitless quantity.

- Meters/meter, μinch/inch, and so on, for strain, reduce to a unitless quantity.

- Ω/square for sheet resistivity reduces to Ω.

- $\Omega \cdot cm^2/cm$ for bulk resistivity reduces to $\Omega \cdot cm$.

6. In typewritten work, substitute u for μ (adding the "tail" by hand if possible), substitute ohm for Ω, and use underline to indicate italic ($\underline{V}$ for V), if necessary, to eliminate confusion.

7. Unit prefixes may be added in front of the SI unit symbol to avoid excessively large or small numbers and the power-of-ten notation. Prefixes are selected to place the number in the range from 0.1 to 1000.

8. The term "billion" and the practice of separating digits into groups of three with commas should be avoided because of conflicting meanings outside the United States. Use a space between groups of three digits for numbers above 9999. For example, write 98 765 432, not 98,765,432.

Common improper or obsolete usages of quantity and unit symbols are illustrated in the table below.

Not Recommended	Proper Form
$v = 15$ fps	$v = 15$ ft/s
psig = 14.7	$p_g = 14.7$ lb/in^2
$E_p = 117$ VAC	$V_p = 117$ V ac
dB = 12	$\alpha = 12$ dB
$I = 1.5$ A$_{rms}$	$I_{rms} = 1.5$ A
$f = 60$ cps	$f = 60$ Hz
$f = 16$ Kc	$f = 16$ kHz
$G = 50$ m$\mu\mho$	$G = 50$ nS
$C = 8$ MFD	$C = 8$ μF
$C = 47$ $\mu\mu$F	$C = 47$ pF

4.2 QUANTITY, UNIT, AND UNIT-PREFIX SYMBOLS

Base units. SI is based on the independently defined units given in the table below.

Quantity	Unit Name	Unit Symbol
Length, l	meter	m
Mass, m	kilogram	kg
Time, t	second	s
Electric current, I	ampere	A
Temperature, T	kelvin	K
Amount of substance	mole	mol
Luminous intensity, I_v	candela	cd
Plane angle, θ	radian	rad
Solid angle, Ω	steradian	sr

• Kilogram is the basic unit of mass. *Kilo* is not regarded here as a prefix. However, additional prefixes should be reduced. For example, microkilogram should be expressed as milligram.

• Kelvin (K), not degree kelvin nor °K, is the basic unit of temperature. However the use of the word *degree* and the symbol (°) are to be continued with °C, °F, and °R.

• A *mole* of any substance contains 6.02257×10^{23} atoms, molecules, ions, or radicals of that substance.

• A *candela* is slightly less than the illumination given by one standard candle (candlepower), and is defined as the amount of light given off by solidifying platinum (1772 °C) through a hole 1 cm² in area.

• A radian is an angle constructed from the center of a circle such that the arc length equals the radius. There are 2π radians in a circle. A steradian is an angle constructed in a sphere such that the surface area equals the radius squared. There are 4π steradians in a sphere.

Derived units that have been given special names are listed in the table on the following page. All SI measurements are to be expressed by combinations of base and derived units.

Derived SI Units

Quantity	Unit Name	Unit Symbol	Formula	SI Base Units
Force, F	newton	N	kg·m/s^2	kg·m/s^2
Pressure, stress, p	pascal	Pa	N/m^2	kg/m·s^2
Energy, work, W	joule	J	N·m	kg·m^2/s^2
Power, P	watt	W	J/s	kg·m^2/s^3
Charge, Q	coulomb	C	A·s	A·s
EMF, V	volt	V	W/A	kg·m^2/A·s^3
Resistance, R	ohm	Ω	V/A	kg·m^2/A^2·s^3
Conductance, G	siemens	S	A/V	A^2·s^3/kg·m^2
Capacitance, C	farad	F	C/V	kg·m^2/s^2
Inductance, L	henry	H	Wb/A	kg·m^2/A^2·s^2
Magnetic flux, φ	weber	Wb	V·s	kg·m^2/A·s^2
Flux density, B	tesla	T	Wb/m^2	kg/A·s^2
Frequency, f	hertz	Hz	1/s	s^{-1}
Luminous flux, φ_v	lumen	lm	cd·sr	cd·sr
Illumination, E_v	lux	lx	lm/m^2	cd·sr/m^2

Magnetic quantities and units are less widely understood than their electrical counterparts, perhaps because magnetic measuring instruments are less common. The table below lists comparable units as an aid to conceptualization.

Electrical Units			**Magnetic Units**		
Quantity	Symbol	Unit	Quantity	Symbol	Unit
EMF	V	V	Magnetomotive force	F_m	A
Field strength	E	V/m	Magnetization	H	A/m
Current	I	A	Magnetic flux	φ	Wb
Current density	J	A/m^2	Flux density	B	T
Resistance	R	Ω	Reluctance	R_m	H^{-1}
Resistivity	ρ	Ω·m	Reluctivity	v	m/H
Conductance	G	S	Permeance	P_m	H
Conductivity	γ	S/m	Permeability	μ	H/m
Relative conductivity	—	—	Relative permeability	μ_r	— (numeric)

Magnetic "Hand" Rules

Field about a wire. Grasp a wire with the right hand, the thumb pointing in the direction of conventional current (positive to negative). The fingers curl around the wire in the direction of the magnetic lines of force (N to S).

Solenoid magnetic polarity. Grasp the coil with the right hand, the fingers curling around in the direction of conventional current. The thumb points to the north pole of the solenoid.

Direction of induced current. Point the index finger of the right hand in the direction of lines of force (N to S) and the thumb in the direction of motion of the conductor. The middle finger (bent toward the palm) points in the direction of induced conventional current.

Force on a moving charge. Point the index finger of the *left* hand in the direction of lines of force (N to S), and the middle finger (bent) in the direction of conventional current. The thumb points in the direction of the force on the charge.

Light is normally measured in the *photometric* system, which includes only the portion of the electromagnetic spectrum visible to the human eye. The *radiometric* system includes all wavelengths of electromagnetic radiation. The two systems are compared in thte table below.

Photometric vs. Radiometric Quantity and Unit Symbols

Quantity	Photometric Symbol	Unit	Comment	Radiometric Symbol	Unit
Source intensity	I_v	cd	cd = lm/sr	I_e	W/sr
Luminous flux (power)	ϕ_v	lm	lm = cd·sr	ϕ_e	W
Illumination (irradiance)	E_v	lx	lx = lm/m^2	E_e	W/m^2
Efficacy	K	lm/W	visible light / total power	—	—

SI quantity and unit symbols (other than magnetic and light) are given in the table below. Symbols in parentheses are reserve quantity symbols to be used only to avoid conflicting symbol meanings.

Additional Quantity and SI Unit Symbols

Quantity	Symbol	Unit
Spatial		
Plane angle	$\theta, \phi, \alpha, \beta$	rad
Solid angle	$\Omega, (\omega)$	sr
Length	l	m
Path length	s	m
Thickness	$d. \delta$	m
Radius	r	m
Diameter	d	m
Area	$A, (S)$	m^2
Volume	V, v	m^3
Time	t	s
Velocity	v	m/s
Angular velocity	ω	rad/s
Acceleration	a	m/s^2
(of free fall)	g	m/s^2
Angular acceleration	α	rad/s^2
Mechanical		
Force	F	N
Weight	W	N
Mass	m	kg
Density	ρ	kg/m^3
Pressure	p	Pa
Momentum	p	kg·m/s
Torque	$T, (M)$	N·m
Rotational inertia	I, J	$kg·m^2$
Work	W	J
Energy	E, W	J
Efficiency	η	numerical
Stress	σ	N/m^2
Strain	ε	numerical

Quantity	Symbol	Unit
Thermal		
Temperature, Celsius	$t, (\theta)$	°C
Temperature, absolute	$T, (\Theta)$	K
Heat energy	Q	J
Heat flow rate	$\Phi, (q)$	W
Thermal resistance	R_θ	K/W
Thermal resistivity	ρ_θ	m·K/W
Heat capacity	C_θ	J/K
Specific heat	c	J/K·kg
Electrical		
Charge	Q	C
Field strength	$E, (K)$	V/m
Electromotive force	V, E	V
Current	I	A
Current density	$J, (s)$	A/m^2
Resistance	R	Ω
Resistivity, volume	ρ	Ω·m
Conductance, $1/R$	G	S
Conductivity	γ, σ	S/m
Reactance	X	Ω
Susceptance, $-1/X$	B	S
Impedance	Z	Ω
Admittance, $1/Z$	Y	S
Characteristic impedance	Z_0	Ω
Transadmittance	y_{xx}	S
Mutual conductance	g_m	S
Amplification factor	μ	numerical
Quality factor, X_S/R_S	Q	numerical
Dissipation factor, $1/Q$	D	numerical
Phase angle	θ, ϕ	rad
Power	P	W
Reactive "power"	$Q, (P_q)$	var
Apparent "power"	$S, (P_s)$	V·A
Power factor	$\cos \phi, F_p$	numerical

Quantity	Symbol	Unit
Period	T	s
Time constant	$\tau, (T)$	s
Frequency	$f, (\nu)$	Hz
Angular frequency	ω	rad/s
Resonant frequency	f_r	Hz
Critical frequency	f_c	Hz
Wavelength	λ	m
Rise time (10%—90%)	t_r	s
Fall time (90%—10%)	t_f	s
Duty factor	D	numerical
Duration of signal element	τ	s
Signaling speed (baud)	$1/t$	Bd
Bandwidth	B	Hz
Noise figure	F	numerical
Amplification (voltage gain)	A_v	numerical
Amplification (current gain)	A_i	numerical
Gain, power	G	numerical
Feedback ratio	$\beta\ (B)$	numerical
Attenuation	α	numerical

SI unit prefixes are given in the table below. Centi, deci, deka, and hecto are to be avoided where possible. Pronounce giga as in *jig*, kilo as in *kill*, nano as in *Nancy*, pico as in *peek*, and peta as in *pet*.

SI Unit Prefixes

Factor	Name	Symbol	Factor	Name	Symbol
10^1	deka	da	10^{-1}	deci	d
10^2	hecto	h	10^{-2}	centi	c
10^3	kilo	k	10^{-3}	milli	m
10^6	mega	M	10^{-6}	micro	μ
10^9	giga	G	10^{-9}	nano	n
10^{12}	tera	T	10^{-12}	pico	p
10^{15}	peta	P	10^{-15}	femto	f
10^{18}	exa	E	10^{-18}	atto	a

4.3 UNIT CONVERSIONS

Conversion factors for converting various units to other commensurate units are given in the table that begins on the following page. Conversions are listed alphabetically according to the general name of the quantity involved— *acceleration* first, *volume* last. A list of miscellaneous conversions, plus a table of Fahrenheit-to-Celsius conversions appears at the end of the main table.

- The first conversion given for each unit is to the basic SI (system international) unit in the mksA (meter-kilogram-second-ampere) system. This is the internationally preferred unit, which should be used wherever possible.

- Boldface conversion numbers are exact. Other factors are given to four-digit precision in the interest of practicality.

- If the desired conversion is not listed, try looking up the reverse conversion. To convert from the unit on the right to the unit listed at the left, *divide* by the factor given.

- If a direct conversion does not appear in the table, multiply by the factor given with the known unit and divide by the factor given with the unknown unit. The process may be represented as *unit A → SI → unit B*. As an example, we convert 40 biblical cubits to feet:

$$40 \text{ cubits} \times 0.4437 \text{ m/cubit} \div 0.3048 \text{ m/ft}$$

$$= 58.23 \text{ ft}$$

- To convert compound units not listed, convert each unit separately. All units except one will have the numerical coefficient 1. As an example, we convert a fuel efficiency of 23 km/liter to miles per gallon.

$$\frac{23 \text{ km}}{1 \text{ L}} = \frac{23 \times 1000 \text{ m}}{1 \times 10^{-3} \text{ m}^3} = \frac{(2300 \div 1609) \text{ mi}}{(10^{-3} \div 3.785 \times 10^{-3}) \text{ gal}}$$

(given units) *(SI units)*   *(convert to desired units)*

$$= 54.1 \text{ mi/gal}$$

Unit Conversions

Boldface numbers are exact.   Others accurate to 4 digits.

To convert from	to	Multiply by
Acceleration		
ft/s^2	m/s^2	**0.3048**
	g (earth grav)	0.03108
	in./s^2	**12**
earth gravity, g	m/s^2	9.807
	ft/s^2	32.17
	in./s^2	386.0
in./sec^2	m/s^2	**0.0254**
	ft/s^2	0.08333
	g (earth grav)	2.590×10^{-3}
m/sec^2	ft/s^2	3.281
	g (earth grav)	0.1020
	in./s^2	39.37
Angle		
degree	rad	0.01745
	gradient	1.111
	minute	**60**
	second	**3600**
gradient	rad	0.01571
	degree	**0.9**
	minute	**54**
	second	**3240**
minute	rad	2.909×10^{-4}
	degree	0.01666
	gradient	0.01852
	second	**60**
radian	degree	57.296
	gradient	63.66
	minute	3438
	second	2.063×10^5
second	rad	4.848×10^{-6}
	degree	2.778×10^{-4}
	gradient	3.086×10^{-4}
	minute	0.01666

To convert from	to	Multiply by
Area		
acre	m^2	4047
	ft^2	**43 560**
	hectare	0.4047
	mile^2	1.5625×10^{-3}
	yard^2	4840
ft^2	m^2	0.09290
	acre	2.296×10^{-5}
	hectare	9.291×10^{-6}
	in.^{2}	**144**
	mile^2	3.587×10^{-8}
	yard^2	0.1111
hectare	m^2	**10 000**
	acre	2.471
	ft^2	107 630
	mile^2	3.861×10^{-3}
	yard^2	11 960
mile^2	m^2	2.590×10^6
	acre	**640**
	ft^2	**2.78784×10^6**
	hectare	259.0
	yard^2	**3.0976×10^6**
yard^2	m^2	0.8361
	acre	2.066×10^{-4}
	ft^2	**9**
	hectare	8.361×10^{-5}
	mile^2	3.228×10^{-7}
barn	m^2	**10^{-28}**
circular mil	m^2	5.067×10^{-10}
	cm^2	5.067×10^{-6}
	in.^{2}	7.854×10^{-7}
	mm^2	5.067×10^{-4}
cm^2	m^2	**0.0001**
	circular mil	1.974×10^5
	in.^{2}	0.1550
	mm^2	**100**

To convert from	to	Multiply by
in.^{2}	m^2	6.4516×10^{-4}
	cm^2	6.4516
	mm^2	645.16
mm^2	m^2	10^{-6}

Density

g/cm^3	kg/m^3	1000
	lb/ft^3	62.43
	lb/in.^{3}	0.03613
	lb/gal	8.345
kg/m^3	g/cm^3	0.001
	lb/ft^3	0.06243
	lb/in.^{3}	3.613×10^{-5}
	lb/gal	8.345×10^{-3}
lb/ft^3	kg/m^3	16.02
	g/cm^3	0.01602
	lb/in.^{3}	5.788×10^{-4}
	lb/gal	0.1337
lb/in.^{3}	kg/m^3	2.768×10^4
	g/cm^3	27.68
	lb/ft^3	1728
	lb/gal	231.0
lb/gal	kg/m^3	119.8
	g/cm^3	0.1198
	lb/ft^3	7.481
	lb/in.^{3}	4.328×10^{-3}

Electrical and Magnetic

ampere·turn	gilbert	1.257
ampere/meter	oersted	0.01257
coulomb	faraday	1.036×10^{-5}
decibel	neper	0.1151
faraday	coulomb	9.652×10^4
gauss	tesla	10^{-4}
gilbert	ampere·turns	0.7958
maxwell (line)	webber	10^{-8}

To convert from	to	Multiply by
oersted	ampere/meter	79.58
unit pole	webber	1.257×10^{-7}
webber	maxwell	**10^8**
webber	unit pole	7.957×10^6
$\Omega \cdot cm$	$\Omega \cdot m$	**0.01**
$\Omega \cdot cirmil/ft$	$\Omega \cdot m$	1.662×10^{-9}
$\Omega \cdot m$	$\Omega \cdot cirmil/ft$	6.017×10^8
$\Omega \cdot mm^2/m$	$\Omega \cdot m$	**10^{-6}**
Energy		
British Thermal Unit	joule, J	1055
(BTU)	calorie	252.0
	erg	1.055×10^{10}
	ft·lb	777.7
	ft·poundal	25 020
	kilocalorie	0.2520
	kW·hr	2.928×10^{-4}
	W·s	1054
calorie	joule, J	4.184
	BTU	3.968×10^{-3}
	electron volt	2.612×10^{19}
	erg	4.184×10^7
	ft·lb	3.086
	kilocalorie	**0.001**
	kW·hr	1.162×10^{-6}
	W·s	4.184
electron volt	joule, J	1.602×10^{-19}
	calorie	3.828×10^{-20}
	erg	1.602×10^{-12}
	W·s	1.602×10^{-19}
erg	joule, J	**10^{-7}**
	BTU	9.484×10^{-11}
	calorie	2.390×10^{-8}
	electron volt	6.242×10^{11}
	ft·lb	7.375×10^{-8}
	kW·hr	2.778×10^{-14}
	W·s	**10^{-7}**

To convert from	to	Multiply by
ft·lb	joule, J	1.356
	BTU	1.284×10^{-3}
	calorie	0.3236
	erg	1.356×10^{7}
	ft·poundal	32.17
	kilocalorie	3.236×10^{-4}
	kW·hr	3.767×10^{-7}
	W·s	1.356
ft·poundal	joule, J	0.04214
	ft·lb	0.03108
kilocalorie	joule, J	4184
	BTU	3.968
	calorie	**1000**
kW·hr	joule, J	$\mathbf{3.6 \times 10^{6}}$
	BTU	3409
	calorie	8.592×10^{5}
	electron volt	2.247×10^{25}
	erg	$\mathbf{3.6 \times 10^{13}}$
	ft·lb	2.655×10^{6}
	kilocalorie	860.5
Force		
dyne	newton, N	$\mathbf{10^{-5}}$
	kg-force	1.020×10^{-6}
	ounce-force	3.597×10^{-5}
	pound-force	2.248×10^{-6}
	poundal	7.236×10^{-5}
kg-force	newton, N	9.807
	dyne	9.807×10^{5}
	ounce-force	35.28
	pound-force	2.205
	poundal	70.94
ounce-force	newton, N	0.2780
	dyne	2.780×10^{4}
	kg-force	0.02835
	pound-force	**0.0625**
	poundal	2.011

To convert from	to	Multiply by
pound-force	newton, N	4.448
	dyne	4.448×10^5
	kg-force	0.4536
	ounce-force	**16**
	poundal	32.17
poundal	newton, N	0.1382
	dyne	13 820
	kg-force	0.01409
	ounce-force	0.4973
	pound-force	0.03108

Length

angstrom	meter, m	$\mathbf{10^{-10}}$
	inch	3.937×10^{-9}
	microinch	3.937×10^{-3}
	mm	$\mathbf{10^{-7}}$
astronomical unit	m	1.496×10^{11}
	light-year	1.581×10^{-5}
	miles	9.298×10^{7}
	parsec	4.848×10^{-6}
cable	m	219.5
	foot	**720**
chain	m	20.12
	fathom	**11**
	foot	**66**
	furlong	**0.1**
	league	4.167×10^{-3}
	mile	**0.0125**
	rod	**4**
	yard	**22**
cubit, English	m	0.4572
	inch	18
cubit, biblical	m	0.5537
	inch	21.8

To convert from	to	Multiply by
fathom	m	1.829
	foot	**6**
foot	m	**0.3048**
	chain	0.01515
	cubit, English	0.6667
	cubit, biblical	0.5505
	fathom	0.1667
	furlong	1.515×10^{-3}
	hand	**3**
	inch	**12**
	league	6.313×10^{-5}
	mile	1.894×10^{-4}
	mile, nautical	1.646×10^{-4}
	rod	0.06061
	span	1.333
	yard	0.3333
furlong	m	201.2
	foot	**660**
hand	m	**0.1016**
	inch	**4**
	foot	0.3333
inch	m	0.0254
	foot	0.08333
	hand	**0.25**
	mil	**1000**
	pica	6.022 (0r **6**)
	point	72.26 (or **72**)
league	m	4828
	foot	**15 840**
	miles	**3**
league, nautical	m	5556
	foot	**18 230**
	miles, nautical	**3**
light-year	m	9.461×10^{15}
	mile	5.878×10^{12}
	parsec	0.3067

To convert from	to	Multiply by
microinch	m	2.54×10^{-8}
	inch	10^{-6}
	micron	0.0254
	mil	0.001
	mm	2.54×10^{-5}
micron	m	10^{-6}
	inch	3.937×10^{-5}
	microinch	39.37
	mil	0.03937
	mm	0.001
mil	m	2.54×10^{-5}
	inch	0.001
	microinch	1000
	micron	25.4
	mm	0.0254
mile	m	1609
	foot	5280
	furlong	8
	km	1.609
	league	0.3333
	light-year	1.701×10^{-13}
	mile, nautical	0.8689
	rod	320
	yard	1760
mile, nautical	m	1852
	foot	6076
	km	1.852
	league, nautical	0.3333
	mile	1.151
parsec	m	3.084×10^{16}
	light-year	3.261
	mile	1.916×10^{13}

To convert from	to	Multiply by
pica	m	4.218×10^{-3}
	inch	0.1667
	mm	4.218
	point	**12**
point	m	3.515×10^{-4}
	inch	0.01388
	mm	0.3515
	pica	0.8333
rod	m	5.029
	chain	**0.25**
	fathom	**2.75**
	foot	**16.5**
	furlong	**0.025**
	league	1.042×10^{-3}
	mile	3.126×10^{-3}
	yard	**5.5**
yard	m	**0.9144**
	chain	0.04545
	fathom	**0.5**
	foot	**3**
	furlong	4.545×10^{-3}
	inch	**36**
	league	1.894×10^{-4}
	mile	5.683×10^{-4}
	mile, nautical	4.937×10^{-4}
	rod	0.1818
Light (source)		
candle	lumen/steradian	**1**
	candela	**1**
candlepower	lumen	12.57
lumen	candlepower	0.07958
lumen	watt	1.5×10^{-3}
Light (illumination)		
ft·candle	lumen/ft^2	**1**
	lumen/m^2	10.76
	lux	10.76

To convert from	to	Multiply by
lumen/m^2	lux	**1**
	ft·candle	0.09290
lumen/ft^2	ft·lambert	**1**
	ft·candle	**1**
	lumen/m^2	10.76
lux	ft·candle	0.09290
Mass		
carat (jeweler's)	kg	**2 × 10^{-4}**
	grain	3.086
	gram	**0.2**
dram, avp	kg	1.772 × 10^{-3}
	grain	27.34
	gram	1.772
	ounce, avp	**0.0625**
grain	kg	6.480 × 10^{-5}
	carat	0.3240
	dram, avp	0.03657
	ounce, avp	2.29 × 10^{-3}
gram	kg	**10^{-3}**
	carat	**5**
	dram, avp	0.5644
	grain	15.43
	pound	2.205 × 10^{-3}
kg (kilogram)	ounce	35.28
	pound	2.205
	slug	0.06852
	stone	0.1575
	ton, short	1.102 × 10^{-3}
	tonne, metric	**10^{-3}**
ounce, avp	kg	0.02835
	carat	141.8
	dram	**16**
	grain	437.5
	gram	28.35
	ounce, troy or apoth	0.9115
	pound	**0.0625**

To convert from	to	Multiply by
ounce, troy or apoth	ounce, avp	1.097
pennyweight	kg	1.555×10^{-3}
	g	1.555
	ounce	0.05486
	pound	3.429×10^{-3}
pound	kg	0.4536
	ounce	**16**
	poundal	32.17
	stone	0.07143
	slug	0.03108
	ton, long	4.464×10^{-4}
	ton, short	5×10^{-4}
	tonne, metric	4.535×10^{-4}
poundal	kg	0.01410
	pound	0.03108
slug	kg	**14.59**
	pound	32.17
stone	kg	6.350
	pound	**14**
ton, long	kg	1016
	pound	**2240**
ton, short	kg	907.2
	pound	**2000**
tonne, metric	kg	**1000**
	pound	2205

Power

BTU/hr	watt	0.2929
	BTU/min	0.01667
	cal/hr	252.0
	cal/min	4.200
	cal/sec	0.0700
	erg/sec	2.929×10^{6}
	ft·lb/sec	0.2160
	horsepower	3.929×10^{-4}
	ton, refrig	8.333×10^{-5}

To convert from	to	Multiply by
cal/sec	watt	**4.184**
	BTU/hr	14.29
	erg/sec	4.184×10^7
	ft·lb/sec	3.085
	horsepower	5.609×10^{-3}
	ton, refrig	1.191×10^{-3}
erg/sec	watt	**10^{-7}**
ft·lb/sec	watt	1.356
	BTU/hr	4.626
	horsepower	1.818×10^{-3}
	ton, refrig	3.856×10^{-4}
horsepower	watt	746
	BTU/hr	2547
	cal/sec	178.3
	ft·lb/sec	**550**
	ft·lb/min	**33 000**
	ton, refrig	0.2121
ton, refrig	watt	3517
	BTU/hr	**12 000**
	horsepower	4.714
watt	BTU/hr	3.415
	cal/min	14.34
	erg/sec	**10^7**
	ft·lb/sec	0.7375
	ft·lb/min	44.25
	horsepower	1.340×10^{-3}
	ton, refrig	2.846×10^{-4}
Pressure		
atmosphere	pascal, Pa	1.013×10^5
bar	pascal	**10^5**
cm of mercury	pascal	1333
dyne/cm^2	pascal	**0.1**
gram/cm^2	pascal	98.07
	lb/in.^{2}	0.01422

To convert from	to	Multiply by
inch of mercury	pascal	3386
lb/in.^{2} (psi)	pascal	6895
	gram/cm^2	70.31
pascal	cm Hg	7.502×10^{-4}
	gram/.cm^2	0.01020
	lb/in.^{2}	70.32
torr (mm Hg)	pascal	133.3

Temperature

°C → K	$t_K = t_{°C} + 273.15°$
°C → F	$t_F = 1.8\, t_{°C} + 32$
°F → K	$t_{°K} = (t_{°F} + 459.7°) \div 1.8$
°F → °C	$t_C = (t_{°F} - 32°) \div 1.8$
°F → °R	$t_R = t_{°F} + 459.7°$

Thermal capacity

BTU/lb·°F	J/kg·K	4187
cal/kg·K	J/kg·K	4.190
J/kg·K	BTU/lb·°F	2.388×10^{-4}
	cal/kg·K	0.2387

Thermal conductivity

BTU·in./hr·ft^2·°F	J/kg·K	0.1442
J/kg·K	BTU·in./hr·ft^2·°F	6.935

(*R* insulating value is the reciprocal of thermal conductivity in BTU·in./hr·ft^2·°F.)

Torque

dyne·cm	N·m	**10^{-7}**
	kg-force·m	1.020×10^{-8}
	oz-force·in.	1.416×10^{-5}
	lb-force·ft	7.375×10^{-8}
kg-force·m	N·m	9.807
	dyne·cm	9.807×10^{-7}
	oz-force·in.	1389
	lb-force·ft	7.23
lb-force·ft	N·m	1.356
	dyne·cm	1.356×10^{-7}
	kg-force·m	0.1383
	oz-force·in.	**192**

To convert from	to	Multiply by
N·m	dyne·cm	10^7
	kg-force·m	0.1020
	oz-force·in.	141.6
	lb-force·ft	0.7375
oz-force·in.	N·m	7.062×10^{-3}
	dyne·cm	7.062×10^4
	kg-force·m	7.201×10^{-4}
	lb-force·ft	5.208×10^{-3}
Velocity		
c (light)	m/s	2.998×10^8
	ft/s	9.836×10^8
	mi/hr	6.706×10^8
ft/min	m/s	5.080×10^{-3}
	ft/s	0.01667
	in./s	0.2
	km/hr	0.01829
	mi/hr	0.01136
ft/s	m/s	0.3048
	ft/min	60
	in./s	12
	km/hr	1.097
	knot	0.5922
	mach number	9.200×10^{-4}
	mi/hr	0.6818
in./s	m/s	0.0254
	mi/hr	0.05682
km/h	m/s	0.2778
	ft/s	0.9114
	knot	0.5400
	mi/hr	0.6214
knot	m/s	0.5144
	ft/s	1.688
	km/hr	1.852
	mi/hr	1.151

To convert from	to	Multiply by
mach number	m/s	331.3
	ft/s	1087
	km/hr	1193
	mi/hr	741.2
mi/h	m/s	**0.44704**
	ft/s	1.467
	km/h	1.609
	knot	0.8690
mi/s	m/s	1609
	c (light)	5.367×10^{-6}
	km/s	1.609
	mi/min	**60**
Volume		
acre·foot	m^3	1233
	barrel, 42-g oil	7758
	ft^3	43 560
	gallon	3.259×10^5
	yd^3	1613
barrel, 42-g oil	m^3	0.1590
	ft^3	5.614
	gallon	**42**
	liter	159.0
board·foot	m^3	2.360×10^{-3}
	ft^3	0.08333
bushel	m^3	0.3524
	ft^3	1.244
	liter	35.24
	peck	**4**
	quart, dry	**32**
	yd^3	0.04607
cord (wood)	m^3	3.625
	ft^3	**128**

To convert from	to	Multiply by
cup	m^3	2.366×10^{-4}
	fl oz	**8**
	gallon	**0.0625**
	gill	**2**
	liter	0.2366
	pint	**0.5**
	quart	**0.25**
	tablespoon	**16**
	teaspoon	**48**
ft^3	m^3	0.02832
	acre·ft	2.296×10^{-5}
	barrel, 42-g oil	0.1781
	board·ft	**12**
	bushel	0.8036
	cord	7.812×10^{-3}
	cup	119.7
	gallon	7.481
	in.^{3}	**1728**
	liter	28.32
	peck	3.215
	perch (stone)	0.04040
	yard^3	0.03704
gallon, US	m^3	3.785×10^{-3}
	acre·ft	3.069×10^{-6}
	bushel	0.1074
	barrel, 42-g oil	0.02381
	cup	**16**
	ft^3	0.1337
	liter	3.785
	peck	0.4296
	pint	**8**
	quart	**4**
	yard^3	4.950×10^{-3}
gallon, Brit.	m^3	4.546×10^{-3}
	gallon, US	1.201

To convert from	to	Multiply by
in.^{3}	m^3	1.639×10^{-5}
	cup	0.06927
	gallon	4.333×10^{-3}
	liter	0.01639
	ounce, fl	0.5541
jigger	ounce, fl	**1.5**
liter (L)	m^3	**10^{-3}**
	barrel, 42-g oil	6.289×10^{-3}
	bushel	0.02838
	cup	4.227
	gallon	0.2642
	in.^{3}	61.01
	ounce, fl	33.82
	pint	2.113
	quart	1.057
	tablespoon	67.61
m^3 (cubic meter)	acre·ft	8.110×10^{-4}
	barrel, 42-g oil	6.289
	board-ft	423.7
	bushel	28.38
	cord	0.2759
	cup	4227
	ft^3	35.31
	gallon	264.2
	liter	**1000**
	perch (stone)	1.427
	yard^3	1.308
milliliter (ml)	m^3	**10^{-6}**
	inch^3	0.06101
	liter	**0.001**
	ounce, fl	0.03382
	teaspoon	0.2029
ounce, fluid	m^3	2.957×10^{-5}
	cup	**0.125**
	gallon	7.812×10^{-3}
	in.^{3}	1.804

To convert from	to	Multiply by
	liter	0.02957
	pint	**0.0625**
	quart	**0.03125**
	tablespoon	**2**
	teaspoon	**6**
peck, US, dry	m^3	8.810×10^{-3}
	barrel, 42-g oil	0.05541
	bushel	**0.25**
	cup, liq	37.24
	ft^3	0.3111
	gallon, liq	2.327
	liter	8.810
	quart, dry	**8**
	quart, liq	9.309
perch (stone)	m^3	0.7004
	ft^3	24.75
pint, liquid	m^3	4.732×10^{-4}
	cup	**2**
	gallon	**0.125**
	in.^{3}	28.87
	liter	0.4732
	ounce, fl	**16**
	quart	**0.5**
	tablespoon	**32**
	teaspoon	**96**
quart, dry	m^3	1.101×10^{-3}
quart, liquid	m^3	9.464×10^{-4}
	cup	**4**
	gallon	**0.25**
	in.^{3}	57.74
	liter	0.9464
	ounce	**32**
	pint	**2**
	quart, dry	0.8594
	tablespoon	**64**
	teaspoon	**192**

To convert from	to	Multiply by
tablespoon	m^3	1.479×10^{-5}
	cup	**0.0625**
	gallon	3.906×10^{-3}
	in.^{3}	0.9024
	liter	0.01479
	ounce, fl	**0.5**
	pint	**0.03125**
	quart	0.01562
	teaspoon	**3**
teaspoon	m^3	4.929×10^{-6}
	cup	0.02083
	in.^{3}	0.3008
	milliliter	4.929
	ounce, fl	0.1667
	pint	0.01042
	tablespoon	0.3333
yd^3	m^3	0.7646
	acre·ft	6.199×10^{-4}
	barrel, 42-g oil	4.809
	bushel	21.70
	cord	0.2109
	ft^3	**27**
	gallon	202.0
	inch^3	**46 656**
	liter	764.6
	peck	86.79
	quart, liq	807.9

Miscellaneous Conversions

Fuel consumption:	1 mi/gal = 0.425 km/liter
	1 mi/gal = 222.2 liter/100 km

To convert mi/gal and liters/100 km, as used in Europe:

$$\text{mi/gal} \times \text{liters/100 km} = 235.3$$

One carat of gold = 41.67 mg/g. Thus 24 kt = pure gold.

Fahrenheit - Celsius Conversions

°C	°F	°C	°F	°C	°F
−100	−148	0	32	36	97
−90	−130	1	34	37	99
−80	−112	2	36	38	100
−70	−94	3	37	39	102
−60	−76	4	39	40	104
−55	−67	5	41	41	106
−50	−58	6	43	42	108
−45	−49	7	45	43	109
−40	−40	8	46	44	111
−35	−31	9	48	45	113
−30	−22	10	50	46	115
−25	−13	11	52	47	117
−24	−11	12	54	48	118
−23	−9	13	55	49	120
−22	−8	14	57	50	122
−21	−6	15	59	55	131
−20	−4	16	61	60	140
−19	−2	17	63	65	149
−18	−0.4	18	64	70	158
−17	1	19	66	75	167
−16	3	20	68	80	176
−15	5	21	70	85	185
−14	7	22	72	90	194
−13	9	23	73	95	203
−12	10	24	75	100	212
−11	12	25	77	125	257
−10	14	26	79	150	302
−9	16	27	81	175	347
−8	18	28	82	200	392
−7	19	29	84	225	437
−6	21	30	86	250	482
−5	23	31	88	275	527
−4	25	32	90	300	572
−3	27	33	91	325	617
−2	28	34	93	350	662
−1	30	35	95	400	752

4.4 PHYSICAL PROPERTIES AND CONSTANTS

Selected physical constants

Length of a 1-s pendulum at sea level	$= 0.994$ m
Solar radiation intensity at earth distance	$= 1.35$ kW/m^2
Mean radius of earth	$= 6370$ km
Mass of earth	$= 5.983 \times 10^{24}$ kg
Earth gravity (g)	$= 9.81$ m/s^2
Gravitational constant (G)	$= 6.670 \times 10^{-11}$ N·m^2/kg^2
Standard atmospheric pressure	$= 1.013 \times 10^5$ N/m^2
Boltzmann's constant (k)	$= 1.381 \times 10^{-23}$ J/K
Planck's constant (h)	$= 6.626 \times 10^{-34}$ J·s
Avogadro's number (N_A)	$= 6.023 \times 10^{23}$ atoms/mole
Permeability of vacuum (μ_0)	$= 4\pi \times 10^{-7}$ H/m
Permittivity of vacuum (ε)	$= 8.854 \times 10^{-12}$ F/m
Electron charge (e)	$= 1.602 \times 10^{-19}$ C
Charge-to-mass ratio of the electron	$= 1.759 \times 10^{11}$ C/kg
Atomic mass unit (C^{12} = 12 amu)	$= 1.6605 \times 10^{-27}$ kg
Electron mass	$= 5.486 \times 10^{-4}$ amu
Proton mass	$= 1.00728$ amu
Neutron mass	$= 1.00867$ amu
Alpha particle mass	$= 4.00151$ amu
Diameter of hydrogen atom	$= 1.06 \times 10^{-10}$ m
Velocity of electromagnetic waves (c_0)	$= 2.998 \times 10^8$ m/s
Velocity of sound in air (14.7 lb/in.^{2}, 0 °C)	$= 331.4$ m/s
Velocity of sound in water	$= 1420$ m/s
Velocity of sound in steel	$= 5103$ m/s

Sound intensity is measured in decibels, with 10^{-12} W/m^2 defined as 0 dB. The lower limit of audibility varies with individuals, but is normally within a few dB of 0 dB. The ear is most sensitive at about 3 kHz. Sensitivity at 300 Hz is down by about 20 dB.

15 dB	whisper	70 dB	truck engine
30 dB	office background	85 dB	jackhammer
55 dB	average conversation	110 dB	jet engine

167

Air pressure (atm) vs. elevation (km)

(km)	0	0.5	1.0	2.0	3.0	5.0	10	20
(atm)	1.00	0.94	0.89	0.78	0.69	0.54	0.26	0.056

Coefficients of sliding friction, $F_{parallel} / F_{normal}$
for selected materials   (numerical)

Rubber on dry pavement	0.75
Rubber on wet pavement	0.50
Wood on hardwood floor	0.25
Steel on steel (dry)	0.20
Steel on steel (oiled)	0.06
Steel runners on ice	0.04

Density at 0 °C and standard pressure
for selected materials (kg/m³)

Gold	19 300	Mercury	13 600
Lead	11 400	Water	1 000
Copper	8 890	Seawater	1 025
Steel	7 830	Ice	910
Aluminum	2 700	Gasoline	690
Magnesium	1 741	Propane	2.02
Cork	224	Carbon dioxide	1.96
Pine wood	430	Air	1.29
Oak wood	750	Helium	0.18
Birch wood	640	Hydrogen	0.09

Young's modulus Y (stress/strain) and
ultimate strength (s_x) for selected metals

Material	Y (N/m² × 10¹⁰)	σ_x (N/m² × 10⁸)
Aluminum	7.0	1.4
Copper	12.5	2.4
Cast iron	9.1	2.9
Mild steel	17.2	4.1
Spring steel	—	13.8
Tungsten	35	41
Magnesium	4.2	1.9

Thermal coefficients of expansion (1/K × 10^{-6})

Linear		Volume	
Aluminum	24	Alcohol	1220
Brass	18	Gasoline	1080
Copper	17	Water at 20 °C	207
Glass	8	Mercury	182
Glass (Pyrex)	3		
Steel	11		
Magnesium	28		

Thermal conductivity (W/m·K)

Silver	415	Sand, concrete	1.8
Copper	381	Glass	0.78
Aluminum	213	Brick	0.72
Steel	50	Wood (fir)	0.11
Lead	35	Cork board	0.037
Ice	2.2	Fiberglass insulation	0.036

Heating values of fuels (J/kg × 10^6)

Hydrogen	142	Fuel oil	45
Natural gas	55	Coal (hard)	33
Gasoline	47	Wood (average)	15
Kerosene	46		

Melting and boiling points at standard pressure (K)

Substance	Melting	Boiling
Helium	0.094	4.3
Nitrogen	63.2	77.1
Carbon dioxide	216.5	213.2
Ammonia	198.7	239.8
Alcohol, ethyl	143.2	350.6
Water	273.2	373.2
Tin	505	2530
Lead	600	1890
Copper	1360	2570
Iron	1810	3260

Heat capacities of various substances (J/kg·K)

Water	4190	Air	1010	Steel	482
Ice	2100	Aluminum	922	Copper	390
Steam	2010	Dry earth	840	Lead	126

Heats of fusion and vaporization (J/kg × 10^6)

Substance	Fusion	Vaporization
Water	0.335	2.26
Ammonia	0.452	1.37
Alcohol	0.104	0.855

Weight of water vapor in saturated air (kg/kg)

°C	kg/kg	°C	kg/kg
10	0.0076	35	0.0370
15	0.0107	40	0.0490
20	0.0149	45	0.0653
25	0.0205	50	0.0868
30	0.0279	55	0.1142

Greek Alphabet

A, α	Alpha (AYL-fa)	N, ν	Nu (New)
B, β	Beta (BAY-tah)	Ξ, ξ	Xi (Zigh)
Γ, γ	Gamma	O, o	Omicron
Δ, δ	Delta	Π, π	Pi
E, ε	Epsilon (EP-sah-lon)	P, ρ	Rho
Z, ζ	Zeta	Σ, σ	Sigma
H, η	Eta	T, τ	Tau (Taow)
Θ, θ	Theta (THAY-tah)	Y, υ	Upsilon
I, ι	Iota	Φ, φ	Phi
K, κ	Kappa	X, χ	Chi
Λ, λ	Lamda	Ψ, ψ	Psi (Sigh)
M, μ	Mu (Mew)	Ω, ω	Omega

5

Standards, Symbols,
and Codes

5.1 FREQUENCY STANDARDS

The equally tempered musical scale for the 88 notes of the piano keyboard is listed in the table below. Each octave contains 12 notes. Each octave is a frequency factor of two above the previous octave, and each note is 2^{-12} or 1.059463 times the frequency of the previous note.

Middle C is 261.6 Hz. A-sharp is the same as B-flat (A# = Bb). Major scales are produced by taking the first, third, fifth, sixth, eighth, tenth, and twelfth notes, starting from the note which names the scale. For example, the scale of A-major contains A, B, C#, D, E, F#, G#, A.

Minor scales are produced by taking the first, third, fourth, sixth, eighth, ninth, and eleventh notes, starting with the note which names the scale. For example, the scale of A-minor contains A, B, C, D, E, F, G, A.

88-Note Keyboard Scale, Hz

A	27.5	55.0	110.0	220.0	440.0	880.0	1760
A#	29.1	58.3	116.5	233.1	466.2	932.3	1865
B	30.9	61.7	123.5	246.9	493.9	987.8	1976
C	32.7	65.4	130.8	261.6	523.3	1047	2093
C#	34.6	69.3	138.6	277.2	554.4	1109	2217
D	36.7	73.4	146.8	293.7	587.3	1175	2349
D#	38.9	77.8	155.6	311.1	622.3	1245	2489
E	41.2	82.4	164.8	329.6	659.3	1319	2637
F	43.7	87.3	174.6	349.2	698.5	1397	2794
F#	46.2	92.5	185.0	370.0	740.0	1480	2960
G	49.0	98.0	196.0	392.0	784.0	1568	3136
G#	51.9	103.8	207.7	415.3	830.6	1661	3322

A 3520	A# 3729	B 3951	C 4186

171

Telephone Touchtone Frequencies (Hz)

1 = 697 + 1209	**7** = 852 + 1209
2 = 697 + 1336	**8** = 852 + 1336
3 = 697 + 1477	**9** = 852 + 1477
4 = 770 + 1209	**0** = 941 + 1336
5 = 770 + 1336	***** = 941 + 1209
6 = 770 + 1477	**#** = 941 + 1477

Broadcast Frequencies

- *AM radio*: 535 to 1705 kHz. 117 channels 10-kHz wide with carriers on multiples of 10 kHz. Tolerance ±10 Hz.

- *European AM broadcast*: 148.5 to 283.5 kHz and 526.5 to 1606.5 kHz

- *FM radio*: 88 to 108 MHz. 100 channels 150 kHz wide with 25-kHz guard bands on each side. Tolerance ±2 kHz. Center frequencies on odd multiples of 0.1 MHz.

- *VHF television*: 54 to 72 MHz (channels 2, 3, and 4), 76 to 88 MHz (channels 5 and 6), 174 to 276 MHz (channels 7 through 13)

VHF Television Frequencies

Channel	Lower limit (MHz)	Channel	Lower limit (MHz)
2	54	8	180
3	60	9	186
4	66	10	192
5	76	11	198
6	82	12	204
7	174	13	210

- *UHF television channels 14 to 69:* Lower limit = $6 (N - 14) + 470$ where N is the UHF channel number.

  TV picture carrier frequency = *lower limit* + 1.25 MHz.

  TV sound carrier frequency = *lower limit* + 5.75 MHz.

- *Regional broadcasting*:

2300 — 2495 kHz
3200 — 3400 kHz
4750 — 4995 kHz

- *International broadcasting*:

5.95 — 6.20 MHz	17.70 — 17.90 MHz
9.50 — 9.775 MHz	21.45 — 21.75 MHz
11.70 —11.975 MHz	25.60 — 26.10 MHz
15.10 —15.45 MHz	

- *Citizen's band (class A) frequencies (MHz)*:
  462.55 — 563.20; 464.75 — 464.95; 465.05 — 466.95

Citizen's band (class D) frequencies[*] (MHz)

Channel	Frequency	Channel	Frequency
1	26.965	21	27.215
2	26.975	22	27.225
3	26.985	23	27.255
4	27.005	24	27.235
5	27.015	25	27.245
6	27.025	26	27.265
7	27.035	27	27.275
8	27.055	28	27.285
9	27.065	29	27.295
10	27.075	30	27.305
11	27.085	31	27.315
12	27.105	32	27.325
13	27.115	33	27.335
14	27.125	34	27.345
15	27.135	35	27.355
16	27.155	36	27.365
17	27.165	37	27.375
18	27.175	38	27.385
19	27.185	39	27.395
20	27.205	40	27.405

* Tolerance ±0.005%

- *Microwave bands (MHz)*:

P	225 – 390	Ku	10 900 – 18 000
L	390 – 1 550	Ka	27 000 – 40 000
S	1 550 – 5 200	Q	36 000 – 46 000
C	3 900 – 6 200	V	46 000 – 75 000
X	5 200 – 10 900	W	75 000 – 110 000

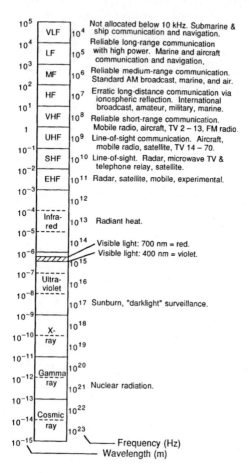

Figure 5.1 The electromagnetic spectrum.

174

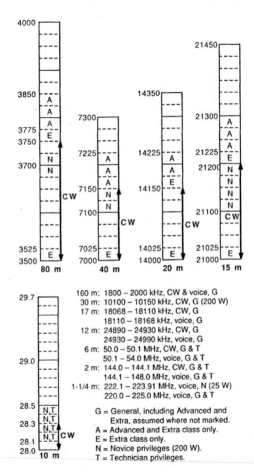

160 m: 1800 – 2000 kHz, CW & voice, G
30 m: 10100 – 10150 kHz, CW, G (200 W)
17 m: 18068 – 18110 kHz, CW, G
        18110 – 18168 kHz, voice, G
12 m: 24890 – 24930 kHz, CW, G
        24930 – 24990 kHz, voice, G
6 m: 50.0 – 50.1 MHz, CW, G & T
        50.1 – 54.0 MHz, voice, G & T
2 m: 144.0 – 144.1 MHz, CW, G & T
        144.1 – 148.0 MHz, voice, G & T
1-1/4 m: 222.1 – 223.91 MHz, voice, N (25 W)
            220.0 – 225.0 MHz, voice, G & T

G = General, including Advanced and
        Extra, assumed where not marked.
A = Advanced and Extra class only.
E = Extra class only.
N = Novice privileges (200 W).
T = Technician privileges.

Figure 5.2   Amateur radio subbands.

Amateur Radio Frequencies

HF and VHF		UHF
160 meters	1.800 – 2.000 MHz	902 – 928 MHz
80 meters	3.500 – 4.000 MHz	1240 – 1300 MHz
40 meters	7.000 – 7.300 MHz	2300 – 2450 MHz
30 meters	10.100 – 10.150 MHz	3300 – 3500 MHz
20 meters	14.100 – 14.350 MHz	5650 – 5925 MHz
17 meters	18.068 – 18.168 MHz	10.0 – 10.5 GHz
15 meters	21.000 – 20.450 MHz	24.0 – 24.25 GHz
12 meters	24.890 – 24.990 MHz	47.0 – 47.2 GHz
10 meters	28.000 – 29.700 MHz	75.5 – 81.0 GHz
6 meters	50.000 – 54.000 MHz	119.98 – 120.02
2 meters	144 - 148 MHz	142 – 149 GHz
1 1/4 meters	222 – 225 MHz	241 – 250 GHz
70 cm	430 – 440 MHz	

5.2 RADIO AND TV BROADCAST STANDARDS

Radio Emission Designations

A0	Unmodulated carrier
A1	Telegraphy (on-off keying)
A2	Tone-modulated telegraphy
A3	Amplitude-modulated telephony
A3A	Single-sideband, reduced carrier
A3B	Double sideband; reduced carrier
A3J	Single sideband, suppressed carrier
A4	Facsimile
A5	Television
F1	Frequency-shift-keying telegraphy/data
F2	Keying of an audio tone in FM
F3	FM telephony
F4	FM facsimile
FM	FM television
P1	Pulsed carrier
P2D	Pulse Amplitude Modulation
P2E	Pulse Width Modulation
P2F	Pulse Position Modulation
P3D	Telephony by Pulse Amplitude Modulation
P3E	Telephony by Pulse Width Modulation
P3F	Telephony by Pulse Position Modulation
P3G	Telephony by Pulse Code Modulation

Radiotelephone Phonetic Alphabet

Alfa	Juliet	Sierra
Bravo	Kilo (KEY-lo)	Tango
Charlie	Lima (LEE-mah)	Uniform
Delta	Mike	Victor
Echo	November	Whiskey
Foxtrot	Oscar	X-ray
Golf	Papa	Yankee
Hotel	Quebec (keh-BECK)	Zulu
India	Romeo	

"S" Units – Radio Reporting System

S1	$0.2\,\mu V$		S4	$1.6\,\mu V$		S7	$13\,\mu V$
S2	$0.4\,\mu V$		S5	$3\,\mu V$		S8	$25\,\mu V$
S3	$0.8\,\mu V$		S6	$6\,\mu V$		S9	$50\,\mu V$

WWV Standard Time Broadcasts.

Stations WWV in Fort Collins, Colorado, and WWVH in Hawaii broadcast continuous time signals on frequencies of 2.5, 5, 10, 15, and 20 MHz. These frequencies are broadcast accurate to one part in 10^{-11} but will be received at somewhat less accuracy because of propagation delays.

Voice announcement of the time is made in the 15 seconds before the beginning of each minute. A man's voice is used at WWV, starting 7.5 seconds before the minute. A woman's voice is used at WWVH, starting 15 seconds before the minute. The beginning of each minute is marked by a 0.8-second 1000 Hz tone from WWV, or a 1200-Hz tone from WWVH. The beginning of each hour is marked by a 0.8-second 1500-Hz tone from both stations.

In general, the first 45 seconds of each minute are filled with standard audio tones: 500 Hz and 600 Hz in alternating minutes. The last 15 seconds contain the voice announcement, or are silent except for a clock tick.

Station identification is given during the 1st and 31st minutes. A standard 440-Hz audio tone (A above middle C) is given during the 2nd minute (WWVH) or 3rd minute (WWV). Special radio propagation and weather alerts are also encoded in the broadcasts.

FM Broadcast Standards , U.S.A.

- Frequency deviation: ±75 kHz maximum.
- Highest modulating frequency: 15 kHz, left + right channels.
- Stereophonic system: Subcarrier (suppressed), 38 kHz with $left - right$ modulation on sidebands. Pilot subcarrier at 19 kHz.

Pre-emphasis of higher audio frequencies is applied in FM broadcast and TV sound transmission. An R-C filter with a critical (–3 dB) frequency of 2120 Hz and a 6-dB/octave rolloff is used at the receiver for de-emphasis. The corresponding R-C time constant is 75 μs, and the amount of pre-emphasis is specified by this time constant.

High-Definition Digital TV Standards

In December of 1996, the FCC approved the use of a new Digital High-Definition TV standard. The analog (NTSC) system is expected to be phased out around 2006. The new system includes wide-screen (16:9 aspect ratio) and six-channel Dolby digital sound. At least six picture formats will be available, including 1280×720 square pixels, and 1920×1080 square pixels. This is two to three times the resolution of the current NTSC system. Both interlaced and non-interlaced formats may be used.

Technically, the Digital HDTV standard is described as having four layers:

The Transmission layer modulates a serial bit stream into a signal that can be transmitted via the standard 6-MHz channel. It uses vestigial-sideband modulation to deliver 19.3 Mbits per second. A pilot tone for rapid signal acquisition and a "training signal" for removing multipath distortion (ghosts) are also included.

The Transport layer packetizes video, audio, and auxiliary data, directing them to the proper receiving components, and enabling the various formats, services, and capabilities of the receiving equipment. HDTV is much more interactive than NTSC televisions, and may be integrated with computers and other appliances.

The Compression layer transforms the raw video and audio data into a set of computer instructions that are executed by the receiver to produce the picture and sound.

The Picture layer consists of raw pixel data, obtained from fairly conventional scan-line and framing processes.

NTSC Television Standards, U.S.A.

- Aspect ratio (width:height) = 4:3
- Channel width: 6.0 MHz
- Video system: Amplitude-modulated; vestigial lower sideband leaving total 4.2 MHz video bandwidth; picture black corresponds to 75% of peak carrier strength, white to 12.5% of carrier strength.
- Sound system: Independent carrier 4.5 MHz above picture carrier; frequency modulated ±25 kHz deviation maximum.
- A *frame* is made up of two *fields* of interlaced horizontal scanning lines. A frame has 525 lines, a field, 262.5 lines.
- Field frequency (vertical sweep frequency): 60 Hz nominal, 59.94 Hz color.
- Line frequency (horizontal sweep frequency): 15 750 Hz nominal, 15 734.26 Hz color.
- Color subcarrier: 3.579545 MHz.

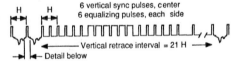

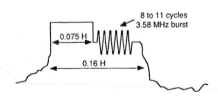

Figure 5.3 Detail of TV vertical and horizontal sync.

International Radio Call-Sign Prefixes

AA-AL	USA	EI-EJ	Ireland
AM-AQ	Spain	EK	Armenia
AP-AW	Pakistan	EL	Liberia
AT-AW	India	EM-EO	Ukraine
AX-AX	Australia	EP-EQ	Iran
AY-AZ	Argentina	ER	Moldova
A2	Botswana	ES	Estonia
A3	Tonga	ET	Ethiopia
A4	Oman	EU-EW	Belarus
A5	Bhutan	EX	Russian Fed
A6	UAEmirates	EY	Tajikistan
A7	Qatar	EZ	Turkmenistan
A8	Liberia	E2	Thailand
A9	Bahrain	F	France
B	China	G	UK Brit, N.Ire
CA-CE	Chile	HA	Hungary
CF-CK	Canada	HB	Switzerland
CL-CM	Cuba	HC-HD	Ecuador
CN	Morocco	HE	Switzerland
CO	Cuba	HF	Poland
CP	Bolivia	HG	Hungary
CQ-CU	Portugal	HH	Haiti
CV-CX	Uruguay	HI	Dominican Rep
CY-CZ	Canada	HJ-HK	Colombia
C2	Nauru	HL	S. Korea
C3	Andorra	HM	N. Korea
C4	Cyprus	HN	Iraq
C5	Gambia	HO-HP	Panama
C6	Bahamas	HQ-HR	Honduras
C7	Meterological	HS	Thailand
C8	Mozambique	HT	Nicaragua
DA-DR	Germany	HU	El Salvador
DS-DT	S. Korea	HV	Vatican City
DU-DZ	Philippines	HW	France
D2-D3	Angola	HZ	Saudi Arabia
D4	Cape Verde	H2	Cyprus
D5	Liberia	H3	Panama
D6	Comoros	H4	Solomon Is
D7-D9	S. Korea	H6-H7	Nicaragua
EA-EH	Spain	H8-H9	Panama

International Radio Call-Sign Prefixes

IA-IZ	Italy	SA-SM	Sweden
JA-JS	Japan	SN-SR	Poland
JT-JV	Mongolia	SSA-SSM	Egypt
JW-JX	Norway	SSN-SSZ	Sudan
JY	Jordan	ST	Sudan
JZ	Indonesia	SU	Egypt
J2	Djibouti	SV-SZ	Greece
J3	Grenada	S2-S3	Bangladesh
J4	Greece	S5	Slovenia
J5	Guinea-Bissau	S6	Singapore
J6	St. Lucia	S7	Seychelles
J7	Domenica	S9	SaoTome, Princ
J8	St.V & Grenadines	TA-TC	Turkey
K	USA	TD	Guatemala
LA-LN	Norway	TE	Costa Rica
LO-LW	Argentina	TF	Iceland
LX	Luxembourg	TG	Guatemala
LY	Lithuania	TH	France
LZ	Bulgaria	TI	Costa Rica
L2-L9	Argentina	TJ	Cameroon
M	UK, Brit & N Ire	TK	France
N	USA	TL	Central Africa
OA-OC	Peru	TM	France
OD	Lebanon	TN	Congo
OE	Austria	TO-TQ	France
OF-OJ	Finland	TR	Gabon
OK-OL	Czech Rep	TS	Tunisia
OM	Slovak Rep	TT	Chad
ON-OT	Belgium	TU	Ivory Coast
OU-OZ	Denmark	TV-TX	France
PA-PI	Netherlands	TY	Benin
PJ	Neth Antilles	TZ	Mali
PK-PO	Indonesia	T2	Tuvalu
PP-PY	Brazil	T3	Kiribati
PZ	Surinam	T4	Cuba
P2	Papua N Guinea	T5	Somalia
P3	Cyprus	T6	Afghanistan
P4	Aruba	T7	San Marino
P5-P9	N. Korea	T9	Bosnia Hrzgov
RA-RZ	Russian Fed	UA-UI	Russian Fed

UJ-UM	Uzbekistan	YO-YR	Romania
UN-UQ	Russian Fed	YS	El Salvador
UR-UZ	Ukraine	YT-YU	Yugoslavia
VA-VG	Canada	YV-YY	Venezuela
VH-VN	Australia	YZ	Yugoslavia
VO	Canada	Y2-Y9	Germany
VP-VS	UK, Brit & N Ire	ZA	Albania
VT-VW	India	ZB-ZJ	UK, Brit N Ire
VX-VY	Canada	ZK-ZM	New Zealand
VZ	Australia	ZN-ZO	UK, Brit N Ire
V2	Antigua, Barbuda	ZP	Paraguay
V3	Belize	ZQ	UK, Brit N Ire
V4	St.Kitts & Nevis	ZR-ZU	South Africa
V5	Namibia	ZV-ZZ	Brazil
V6	Micronesia	Z2	Zimbabwe
V7	Marshall Is.	Z3	Macedonia
V8	Brunei	2A-2Z	UK, Brit N Ire
W	USA	3A	Monaco
XA-XI	Mexico	3B	Mauritius
XJ-XO	Canada	3C	Equator Guinea
XP	Denmark	3DA-3DM	Swaziland
XQ-XR	Chile	3DN-3DZ	Fiji
XS	China	3E-3F	Panama
XT	Burkina Faso	3G	Chile
XU	Cambodia	3H-3U	China
XV	Viet Nam	3V	Tunisia
XW	Laos	3W	Viet Nam
XX	Portugal	3X	Guinea
XY-XZ	Myanmar	3Y	Norway
YA	Afghanistan	3Z	Poland
YB-YH	Indonesia	4A-4C	Mexico
YI	Iraq	4D-4I	Philippines
YJ	Vanauatu	4J-4K	Azerbaijan
YK	Syria	4L	Georgia
YL	Latvia	4M	Venezuela
YM	Turkey	4N-4O	Yugoslavia
YN	Nicaragua	4P-4S	Sri Lanka

International Radio Call-Sign Prefixes

4T	Peru	7O	Yemen
4U	United Nations	7P	Lesotho
4V	Haiti	7Q	Malawi
4X	Israel	7R	Algeria
4Y	Aviation	7S	Sweden
4Z	Israel	7T-7Y	Algeria
5A	Libya	7Z	Saudi Arabia
5B	Cyprus	8A-8I	Indonesia
5C-5G	Morocco	8J-8N	Japan
5H-5I	Tanzania	8O	Botswana
5J-5K	Colombia	8P	Barbados
5L-5M	Liberia	8Q	Maldives
5N-5O	Nigeria	8R	Guyana
5P-5Q	Denmark	8S	Sweden
5R-5S	Madagascar	8T-8Y	India
5T	Mauritania	8Z	Saudi Arabia
5U	Niger	9A	Croatia
5V	Togo	9B-9D	Iran
5W	Western Samoa	9E-9F	Ethiopia
5X	Uganda	9G	Ghana
5Y-5Z	Kenya	9H	Malta
6A-6B	Egypt	9I-9J	Zambia
6C	Syria	9K	Kuwait
6D-6J	Mexico	9L	Sierra Leone
6K-6N	S. Korea	9M	Malaysia
6O	Somalia	9N	Nepal
6P-6S	Pakistan	9O-9T	Zaire
6T-6U	Sudan	9U	Burundi
6V-6W	Senegal	9V	Singapore
6X	Madagascar	9W	Malaysia
6Y	Jamaica	9X	Rwanda
6Z	Liberia	9Y-9Z	Trinidad &
7A-7I	Indonesia		Tobago
7J-7N	Japan		

5.3 COMMUNICATIONS CODES

International Morse (Continental) Code

A ·—	N —·	1 ·————	6 —····
B —···	O ———	2 ··———	7 ——···
C —·—·	P ·——·	3 ···——	8 ———··
D —··	Q ——·—	4 ····—	9 ————·
E ·	R ·—·	5 ·····	0 —————
F ··—·	S ···	Period ·—·—·—	
G ——·	T —	Comma ——··——	
H ····	U ··—	Question ··——··	
I ··	V ···—	Slash bar —··—·	
J ·———	W ·——	Break —···—	
K —·—	X —··—	Error ········	
L ·—··	Y —·——	Wait ·—···	
M ——	Z ——··	Go ahead —·—	

A dash is equal to three dots.
The space between elements is equal to one dot.
The space between letters is equal to one dash.

Old American Morse (Landline) Code

A ·—	N —·	1 ·——·	6 ······
B —···	O ··	2 ··—··	7 ——··
C ···	P ·····	3 ···—·	8 —····
D —··	Q ··—·	4 ····—	9 —··—
E ·	R ···	5 ———	0 —
F ·—·	S ···	Period ··—···	
G ——·	T —	Comma ·—·—	
H ····	U ··—	Question —··—·	
I ··	V ···—	Paragraph ——·—·	
J —·—·	W ·——	Semicolon ··· ··	
K —·—	X ·—··	Exclamation ——··	
L —	Y ·· ··	Colon —·—··	
M ——	Z ··· ·	Quote ··—· —··	

Note the slightly longer space between dots in letters
C, O, R, Y, and Z, and in the semicolon, and quote.

Common Radio "Q" Signals

QRA	name of station	QRX	stand by
QRG	frequency	QRZ	who is calling?
QRK	readability	QSA	signal strength
QRL	busy; in use	QSB	fading
QRM	interference	QSK	break-in keying
QRN	static	QSL	confirm message
QRO	higher power	QSM	repeat message
QRP	lower power	QSO	contact
QRQ	send faster	QSV	send Vs for test
QRS	send slower	QSX	listen for station
QRT	stop sending	QSY	change frequency
QRU	no message	QSZ	send words twice
QRV	ready for message	QTH	location

Baudot 5-Level Teletype Code (1 = hole)

Letter	Figure	54321	Letter	Figure	54321
A	—	00011	Q	1	10111
B	?	11001	R	4	01010
C	:	01110	S	Bell	00101
D	$	01001	T	5	10000
E	3	00001	U	7	00111
F	!	01101	V	;	11110
G	&	11010	W	2	10011
H	#	10100	X	/	11101
I	8	00110	Y	6	10101
J	'	01011	Z	"	10001
K	(	01111	Carriage return		01000
L	)	10010	Line Feed		00010
M	.	11100	Letters		11111
N	,	01100	Figures		11011
O	9	11000	SPACE		00100
P	0	10110	Sprocket holes between 2 & 3		

Police Radio "Ten" Code

10-1	Unable to copy; change location
10-2	Signals good
10-3	Stop transmitting
10-4	Acknowledge
10-5	Relay message
10-6	Busy; stand by unless urgent
10-7	Out of service
10-8	In service
10-9	Repeat message
10-10	Fight in progress
10-11	Dog case
10-12	Stand by
10-13	Weather and road report
10-14	Report of prowler
10-15	Civil disturbance
10-16	Domestic trouble
10-17	Meet complainant
10-18	Complete assignment quickly
10-19	Return to . . .
10-20	Location
10-21	Call . . . by telephone
10-22	Disregard
10-23	Arrived at scene
10-24	Assignment completed
10-25	Report in person to . . .
10-26	Detaining subject; expedite
10-27	Driver's license information
10-28	Vehicle-registration information
10-29	Check records for wanted
10-30	Illegal use of radio
10-31	Crime in progress
10-32	Man with gun
10-33	Emergency
10-34	Riot
10-35	Major crime alert
10-36	Correct time
10-37	Investigate suspicious vehicle
10-38	Stopping suspicious vehicle
10-39	Urgent; use lights and siren
10-40	Silent run; no lights or siren
10-41	Beginning tour of duty
10-42	Ending tour of duty
10-43	Information
10-44	Request permission to leave patrol for . . .

10-45	Animal carcass
10-46	Assist motorist
10-47	Emergency road repairs needed
10-48	Traffic standard needs repairs
10-49	Traffic light out
10-50	Accident
10-51	Wrecker needed
10-52	Ambulance needed
10-53	Road blocked
10-54	Livestock on highway
10-55	Intoxicated driver
10-56	Intoxicated pedestrian
10-57	Hit and run
10-58	Direct traffic
10-59	Convoy or escort
10-60	Squad in vicinity
10-61	Personnel in area
10-62	Reply to message
10-63	Prepare to make written copy
10-64	Message for local delivery
10-65	Net message assignment
10-66	Message cancellation
10-67	Clear to read net message
10-68	Dispatch information
10-69	Message received
10-70	Fire alarm
10-71	Advise nature of fire
10-72	Report progress on fire
10-73	Smoke report
10-74	Negative
10-75	In contact with
10-76	En route
10-77	ETA (estimated time of arrival)
10-78	Need assistance
10-79	Notify coroner
10-82	Reserve lodging
10-84	Are you going to meet . . . ?
10-85	Will be late
10-87	Pick up checks for distribution
10-88	Advise telephone number to contact
10-90	Bank alarm
10-91	Unnecessary use of radio
10-93	Blockade
10-94	Drag racing
10-96	Mental subject
10-98	Prison or jail break
10-99	Records indicate wanted or stolen

5.4 MECHANICAL HARDWARE STANDARDS

Sheet Metal Gauges (Decimal Inches)

Gauge No.	American or B&S Aluminum, Copper, Brass	US Standard Iron, Steel, Nickel	Birmingham or Stubs Tubes; by some mfgrs Copper & Brass
10	0.1019	0.140625	0.134
11	0.09074	0.125	0.120
12	0.08081	0.109375	0.109
13	0.07196	0.09375	0.095
14	0.06408	0.078125	0.083
15	0.05707	0.0703125	0.072
16	0.05082	0.0625	0.065
17	0.04526	0.05625	0.058
18	0.04030	0.0500	0.049
19	0.03589	0.04375	0.042
20	0.03196	0.0375	0.035
21	0.02846	0.034375	0.032
22	0.02535	0.03125	0.028
23	0.02257	0.028125	0.025
24	0.02010	0.0250	0.022
25	0.01790	0.021875	0.020
26	0.01594	0.01875	0.018
27	0.01420	0.017188	0.016
28	0.01264	0.015625	0.014

Machine-Screw Tap and Clearance Drill Sizes

Screw	Tap	Clear	Screw	Tap	Clear
2-56	50	42	6-32	36	25
2-64	50	42	6-40	33	25
3-48	47	36	8-32	29	16
3-56	45	36	8-36	29	16
4-40	43	31	10-24	25	13/64
4-48	42	31	10-32	21	13/64
5-40	38	29	12-24	16	7/32
5-44	37	29	12-28	14	7/32

Decimal Equivalents of Standard Drill Sizes

Size	Inches	Size	Inches	Size	Inches
70	0.0280	44	0.0860	19	0.1660
69	0.0292	43	0.0890	18	0.1695
68	0.0310	42	0.0935	11/64	0.1709
1/32	0.0313	3/32	0.0938	17	0.1730
67	0.0320	41	0.0960	16	0.1770
66	0.0330	40	0.0980	15	0.1800
65	0.0350	39	0.0995	14	0.1820
64	0.0360	38	0.1015	13	0.1850
63	0.0370	37	0.1040	3/16	0.1875
62	0.0380	36	0.1065	12	0.1890
61	0.0390	7/64	0.1094	11	0.1910
60	0.0400	35	0.1100	10	0.1935
59	0.0410	34	0.1110	9	0.1960
58	0.0420	33	0.1130	8	0.1990
57	0.0430	32	0.1160	7	0.2010
56	0.0465	31	0.1200	13/64	0.2031
3/64	0.0469	1/8	0.1250	6	0.2040
55	0.0520	30	0.1285	5	0.2055
54	0.0550	29	0.1360	4	0.2090
53	0.0595	28	0.1405	3	0.2130
1/16	0.0625	9/64	0.1406	7/32	0.2188
52	0.0635	27	0.1440	2	0.2210
51	0.0670	26	0.1470	1	0.2280
50	0.0700	25	0.1495	A	0.2340
49	0.0730	24	0.1520	15/64	0.2344
48	0.0760	23	0.1540	B	0.2380
5/64	0.0781	5/32	0.1562	C	0.2420
47	0.0785	22	0.1570	D	0.2460
46	0.0810	21	0.1590	1/4, E	0.2500
45	0.0820	20	0.1610		

5.5 STANDARD SCHEMATIC SYMBOLS

Resistor, fixed.

Resistor with adjustable contact; potentiometer.

Resistor; 2-terminal adjustable; rheostat.

Resistor with preset adjustment; trimmer.

Temperature-dependent resistor; thermistor.

Resistor sensitive to nonionizing radiation (light or infrared).

Resistor sensitive to ionizing radiation: x-ray, gamma ray, alpha, beta particle, etc.

Capacitor, IEC standard symbol. Plate length 3 to 5 times plate spacing.

Capacitor, popular symbol. Curved line represents outside plate or negative if polarized. Plus sign used only with polarized types.

Capacitor; variable. Left plate movable. Alternative symbols given. Left symbol most common.

Capacitor with preset adjust; trimmer.

Capacitors; variable, mechanically ganged.

Capacitor; variable differential. One side increases as other side decreases.

Capacitor; split stator. Both sides increase together.

Inductor; winding; coil. General and air-core.

Inductor, magnetic core.

Inductor, powdered-iron or ferrite core.

Transformer; general and air-core. Dots show phase.

Transformer, adjustable coupling.

Transformer; magnetic core, nonsaturating.

Transformer with electrostatic shield between windings.

Transformer with saturable magnetic core.

Wires connected.

Form to be avoided. Offset lines as at left.

Wires crossing,
not connected;
preferred form.

Obsolete for wires
crossing, not connected.
Do not use.

Shielded conductors
(2-wire shown).

Coaxial cable.

Connector, engaged.
3-wire male plug (left)
and female receptacle
(right) shown.

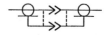

Connectors, coaxial.

Connector, male plug,
2-wire, nonpolarized.

Connector, female
receptacle, 3-wire.

Common connection
between identical letters.

Chassis "ground," or
common connection
in instruments without
conductive chassis.

Earth ground or metal
vehicle frame.

Antennas; general (left),
and dipole (right).

Loop antenna.

Antenna counterpoise.

Fuse, general.

Circuit breaker.

Thermal cutout switch.

Single-throw switch.

Double-throw switch.

Rotary switch.

Multiposition switch.

Pushbutton switch;
normally open (left),
normally closed (right).

Flasher or self-
interrupting switch.

Limit switch, closed by
moving machinery.

Limit switch, opened by
moving machinery.

193

Switch, normally
closed, opens after
time delay.

Switch, normally
open, closes on
rising temperature.

Switch, normally
closed, opens on
rising liquid level.

Switch, normally
open, closes on
increased fluid flow.

Switch, normally
closed, opens on
increased pressure.

Centrifugal switch,
opens on increased
angular speed.

Relay or
solenoid coil.

Relay contacts; normally
open (left) and closed
(right). Make length no
more than 1.5 times
spacing to avoid
confusion with
capacitor symbol.

Alternative contacts.

Battery; single cell (left)
and multicell (right).
Polarity sign optional.

AC source, general.
Oscillator.

Thermocouple.

Piezoelectric crystal.

Magnetic head:
erase (x), record (→),
and read (←) functions.

Microphone.

Loudspeaker.

Headphone; earphone.

Incandescent lamp;
pilot or indicating.

Incandescent lamp;
illuminating.

Lamp, fluorescent,
with heaters.

Cold-cathode glow
lamp; neon lamp.

Meter, general.
Label defines type.

Rotating machine.
Label defines type.

Conventional flow ⟶
Allowed electron flow ⟵

Semiconductor diode.

Regulator, avalanche, or zener diode.

Varactor diode; varicap.

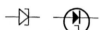

Esaki (tunnel) diode.

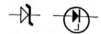

Transistor, NPN.

Transistor, PNP.

Unijunction transistor.

Junction FET, N-channel.
Reverse arrow for
P-channel.

Insulated-gate or
MOS FET, N-channel,
depletion type.

Insulated-gate FET,
P-channel,
enhancement type.

Silicon controlled
rectifier (SCR).

Gate turn-off SCR
(GTO).

Triac.

Diac.

Silicon bilateral switch
(SBS).

Programmable UJT (PUT).
Pins compared with UJT.

Amplifier, general. Signs
indicate inverting and
noninverting inputs.

Integrator, general.

Analog multiplier.

Analog divider.

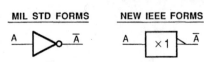

Digital logic inverter (NOT gate).

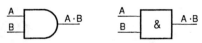

AND gate: All inputs true required for true output.

OR gate: Only one true input required for output true.

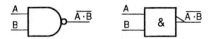

NAND gate: All inputs true required for false output.

NOR gate: Any one input true makes output false.

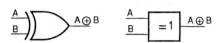

Exclusive OR: Output is true if one input is true,
but not if more than one input is true.

INTERNATIONAL EQUIPMENT MARKINGS

○ ●	Power: ON OFF	**V3**	VHF TV channel
∿	AC voltage only	◿	Increase volume
∿ (with bar)	AC/DC voltage	♪:	Bass
50/60	50–60 Hz power	𝄞	Treble
⚡	High voltage	☼	Brightness
⏚	Ground	◑	Contrast
⚠	Hazard warning	←•→	Horizontal hold
⌀	Phono input	↕	Vertical hold
◎◎	Tape input	⊕	Horizontal linearity
Y	Wire antenna	⊕	Vertical linearity
⊓⊓	Dipole antenna	⊕	
◁	Speaker output	|←→|	Width
⊐	Earphone	I	Height
←	Left channel		
→	Right channel		
▨	Remote control socket		

6

Computer Standards and Codes

6.1 DIGITAL CODES

8-4-2-1 code. In most common use. BCD if 0 through 9; hexadecimal if 0 – F.

0	0000	4	0100	8	1000	C	1100
1	0001	5	0101	9	1001	D	1101
2	0010	6	0110	A	1010	E	1110
3	0011	7	0111	B	1011	F	1111

Excess-3 code. 1's complement of binary gives 9's complement of decimal.

0	0011	3	0110	6	1001	9	1100
1	0100	4	0111	7	1010		
2	0101	5	1000	8	1011		

Gray code. Only one bit changes between counts.

0	0000	4	0110	8	1100	12	1010
1	0001	5	0111	9	1101	13	1011
2	0011	6	0101	10	1111	14	1001
3	0010	7	0100	11	1110	15	1000

Two's complement. A + 2's comp of B = A − B.

0	0000	4	1100	8	1000	12	0100
1	1111	5	1011	9	0111	13	0011
2	1110	6	1010	10	0110	14	0010
3	1101	7	1001	11	0101	15	0001

ASCII Seven-Level Code

Hex	ASCII	Hex	ASCII	Hex	ASCII	Hex	ASCII
00	Null	20	space	40	@	60	`
01	Start head	21	!	41	A	61	a
02	Start text	22	"quote	42	B	62	b
03	End text	23	#	43	C	63	c
04	End trans	24	$	44	D	64	d
05	Enquiry	25	%	45	E	65	e
06	Acknledg	26	&	46	F	66	f
07	Bell	27	'apost	47	G	67	g
08	Backspc	28	(	48	H	68	h
09	Hor tab	29	)	49	I	69	i
0A	Line feed	2A	*	4A	J	6A	j
0B	Vert tab	2B	+	4B	K	6B	k
0C	Form feed	2C	,comma	4C	L	6C	l
0D	Return	2D	-hyphen	4D	M	6D	m
0E	Shift out	2E	.period	4E	N	6E	n
0F	Shift in	2F	/	4F	O	6F	o
10	Link esc.	30	0	50	P	70	p
11	Cntrl 1	31	1	51	Q	71	q
12	Cntrl 2	32	2	52	R	72	r
13	Cntrl 3	33	3	53	S	73	s
14	Cntrl 4	34	4	54	T	74	t
15	Neg ack	35	5	55	U	75	u
16	Sync idle	36	6	56	V	76	v
17	End block	37	7	57	W	77	w
18	Cancel	38	8	58	X	78	x
19	End med'm	39	9	59	Y	79	y
1A	Subst	3A	:colon	5A	Z	7A	z
1B	Escape	3B	;semic	5B	[	7B	{
1C	File sepr	3C	<	5C	\	7C		
1D	Group sep	3D	=	5D	]	7D	}
1E	Record sep	3E	>	5E	^	7E	≈
1F	Unit sepr	3F	?	5F	—dash	7F	Del

Extended Binary-Coded Decimal Interchange Code

0	F0	m	94	I	C9	+	4E
1	F1	n	95	J	D1	!	4F
2	F3	o	96	K	D2	&	50
3	F3	p	97	L	D3	!	5A
4	F4	q	98	M	D4	$	5B
5	F5	r	99	N	D5	*	5C
6	F6	s	A2	O	D6	)	5D
7	F7	t	A3	P	D7	:	5E
8	F8	u	A4	Q	D8	-	60
9	F9	v	A5	R	D9	/	61
a	81	w	A6	S	E2	'	6B
b	82	x	A7	T	E3	%	6C
c	83	y	A8	U	E4	-	6D
d	84	z	A8	V	E5	>	6E
e	85	A	C1	W	E6	?	6F
f	86	B	C2	X	E7	#	7B
g	87	C	C3	Y	E8	@	7C
h	88	D	C4	Z	E9	,	70
i	89	E	C5	¢	4A	=	7E
j	91	F	C6	.	4B	"	7F
k	92	G	C7	<	4C	blank	40
l	93	H	C8	(	4D	linfed	25

Zone Punch	No	Row punch										3 +8	4 +8
		0	1	2	3	4	5	6	7	8	9		
None		0	1	2	3	4	5	6	7	8	9	=	'
12	+		A	B	C	D	E	F	G	H	I	.	)
11	−		J	K	L	M	N	O	P	Q	R	$	*
0			/	S	T	U	V	W	X	Y	Z	,	(

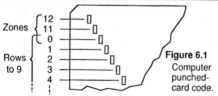

Zones { 12, 11, 0
Rows to 9 { 0, 1, 2, 3, 4, ...

Figure 6.1
Computer punched-card code.

6.2 BOOLEAN ALGEBRA THEOREMS

Symbols: "+" is logic OR; "." is logic AND.

1. $0 \cdot 0 = 0$ $0 + 0 = 0$

2. $0 \cdot 1 = 0$ $0 + 1 = 1$

3. $1 \cdot 1 = 1$ $1 + 1 = 1$

4. $\overline{1} = 0$ $\overline{0} = 1$

5. If $A = 0$, $\overline{A} = 1$ If $A = 1$, $\overline{A} = 0$

6. $A \cdot 0 = 0$ $A + 0 = A$

7. $A \cdot 1 = A$ $A + 1 = 1$

8. $A \cdot A = A$ $A + A = A$

9. $A \cdot \overline{A} = 0$ $A + \overline{A} = 1$

10. $\overline{\overline{A}} = A$ $A = \overline{\overline{A}}$

11. $A \cdot B = B \cdot A$ $A + B = B + A$
    (commutative property)

12. $A \cdot (B \cdot C) = (A \cdot B) \cdot C$ $A + (B + C) = (A + B) + C$
    (associative property)

13. $A(B + C) = AB + AC$ $A + BC = (A + B)(A + C)$
    (distributive property)

14. $A \cdot (A + B) = A$ $A + (A \cdot B) = A$

15. $A \cdot (\overline{A} + B) = A \cdot B$ $A + (\overline{A} \cdot B) = A + B$
    (absorption)

16. $\overline{A \cdot B \cdot C} = \overline{A} + \overline{B} + \overline{C}$ $\overline{A + B + C} = \overline{A} \cdot \overline{B} \cdot \overline{C}$
    (deMorgan's theorem)

6.3 STANDARD COMPUTER CONNECTORS

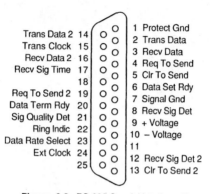

Figure 6.2 RS-232 C serial interface; Pin side of plug, solder side of receptacle.

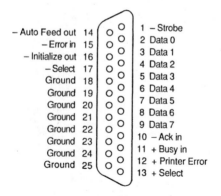

Figure 6.3 DB-25 parallel printer interface; connector side of line plug.

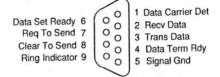

Figure 6.4 Standard DB-9 serial port.

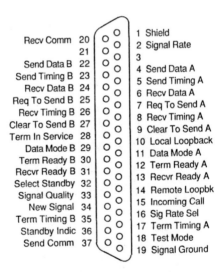

Figure 6.5 RS-449 high-speed serial interface; pin side of plug.

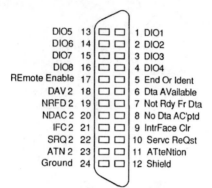

Figure 6.6 IEEE-488 parallel bus (GPIB).

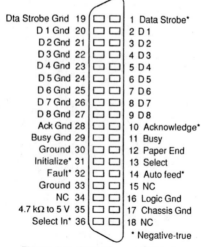

Figure 6.7 Centronics parallel interface.

Ground	B1	A1	I/O Chk
Reset out	B2	A2	D 7
+5 V	B3	A3	D 6
IntReq2	B4	A4	D 5
−5 V	B5	A5	D 4
DmaReq2	B6	A6	D 3
−12 V	B7	A7	D 2
	B8	A8	D 1
+12 V	B9	A9	D 0
Ground	B10	A10	I/O Rdy
Mem Wrt	B11	A11	Adr EN
Mem Read	B12	A12	A19
I/O Wrt	B13	A13	A18
I/O Read	B14	A14	A17
DtaAck3	B15	A15	A16
DtaReq3	B16	A16	A15
DtaAck1	B17	A17	A14
DtaReq3	B18	A18	A13
DtaAck0	B19	A19	A12
Clock	B20	A20	A11
IntReq7	B21	A21	A10
IntReq6	B22	A22	A9
IntReq5	B23	A23	A8
IntReq4	B24	A24	A7
IntReq3	B25	A25	A6
DtaAck2	B26	A26	A5
TermCount	B27	A27	A4
AdrLatchEn	B28	A28	A3
+5 V	B29	A29	A2
Osc	B30	A30	A1
Ground	B31	A31	A0

Figure 6.8 IBM PC and Compatible Bus.

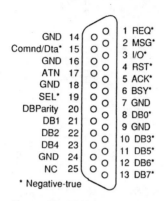

Figure 6.9 SCSI (Small Computer Systems Interface) connector; pin side of plug.

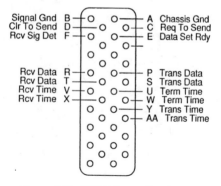

Figure 6.10 V.35 high-speed interface.

The Universal Serial Bus (USB)

The Universal Serial Bus (Figure 6.11, below) permits data exchange between a host computer and a number of peripherals. Individual peripherals may be disconnected and reconnected while the system is in operation.

USB "hubs" are wiring concentrators with a single port connected "upstream" toward the host computer, and multiple ports connected "downstream" toward the peripherals. Upstream and downstream connectors are physically different to avoid misconnections.

The USB cable is four-conductor with a differential signal of 200 mV minimum between two of the lines. The VBus and ground wires provide 5 V to power peripherals.

USB Cable Color Code

+Data	Green	200 mV differential
–Data	White	on twisted pair
VBus	Red	+5 V
Ground	Black	Power ground

Maximum cable length is 5 meters (16 ft.) Maximum data rate is 12 Mbits/second, with a lower speed of 1.5 Mbs available. Data is sent in packets, with a sync field preceding each packet. A clock signal is encoded along with the differential data. *A-series* connectors have four pins in-line. *B-series* connectors have the pins in a four-square arrangement.

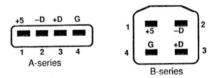

Figure 6.11 Universal Serial Bus.

6.4 COMPUTER-LANGUAGE SUMMARIES

Basic PC DOS Operations

- Boldface characters are those to be typed. Lightface characters are those generated by the computer.

- All commands are assumed to be terminated with a RETURN (↵) key press. Sometimes DOS offers a list of responses, such as Yes or Retry. Type the first letter.

- Uppercase and lowercase letters are equivalent in DOS commands.

ESC   Eliminates an entire line (before ENTER)

CTRL + NumLock   Makes a scrolling screen pause. Hit any key to continue. (CTRL + S may also work.)

CTRL + Scroll Lock/Break   Stops a command in progress.

CTRL + ALT + DEL   Cold boot; restarts computer.

Shift + PrtSc   Prints a copy of the current screen display.

Ctrl + PrtSc   Turns printer echo on / off. Everything typed will appear on the printer as it is entered.

A:>c: Changes from drive A (floppy) to C (hard).

C:>cd   Displays name of current directory.

C:>cd   Changes to the root directory.

C:>cd\school\english   Changes to the *english* subdirectory under *school* subdirectory under root directory.

C:>dir   Displays names of all files on drive C, with size in bytes and time of last change to each file.

C:>dir a:/p   Displays one screen (24 lines) of drive-A directory, then pauses and waits for any key to continue.

C:>dir/w   Displays directory in wide format (5 columns, filenames only, 85 files per screen.)

C:>dir b:>prn   Prints directory of drive B.

C:>ren drftrpt.ext finlrpt.ext   Renames the file *drftrpt* as *finlrpt*. File names may contain up to 8 characters, A – Z, 0 – 9, plus ! @ # $ % ( ) - _ { } and ' The three *extension* letters after the period are optional.

`C:>del drftrpt.ext`   Deletes the file named *drftrpt*.

`C:>erase drftrpt.ext`   Same as *del* command.

`C:>copy c:finalrpt.ext a:`   Copies the file named *finalrpt* from the hard drive to floppy drive A.

`C:>copy finalrpt.ext c:\archive`   Places a copy of *finalrpt* (from the current directory on the hard drive) under the directory *archive* on the hard drive.

`C:>copy c:*.eng a:`   Copies all files from drive C with the extension *eng* to drive A.

`C:>copy a:*.* b:`   Copies all files from drive A to the disk in drive B.

`C:>diskcopy a: b:`   Copies everything from the disk in drive A to the disk in drive B. The old contents of disk B will be erased, and disk B will be formatted if necessary. Requires drives of same disk size.

`C:>diskcopy a: a:`   Makes a copy of a floppy using a single floppy drive.

`C:>format a:`   Erases the disk in drive A and prepares it to receive DOS files. WARNING: **Never format c:** or you'll wipe out the entire hard drive.

`C:>format b: /s`   Formats disk in drive B and adds a copy of DOS so a computer can "boot up" from this disk.

`C:>format a: /4`   Formats a standard (360 K) disk in a high-capacity (1.2 M) drive.

`C:>chkdsk a:`   Reports on disk in drive A: bytes used, number of files, bytes left, sizes of data records, and computer (RAM) memory free. Suffix **/f** may added to fix file errors that may have occurred.

`C:>edit filename.ext`   Version 5.0: full-screen text editor. Files with extensions **.txt** and **.doc** can be viewed.

`C:>mem/c`   Version 5.0: memory-usage information.

`C:>help command`   Version 5.0 information call.

`*.EXE`, `*.COM`   Executable program files. May be started by typing file name without extension.

`*.BAT`   Batch files that are executed sequentially.

`*.TXT`, `*.DOC`   Text files.

PC DOS Instruction Summary

For help with MS-DOS commands, type **HELP** at the DOS prompt, or type the command name followed by **/?** (e.g. **FORMAT /?**) The following is a partial list of DOS 6.2 commands.

append	Make specified directories available to programs running in current directory.
attrib	Display or change file attributes.
call	Call one batch program from another.
cd, dhdir	Display name of, or change, directory.
chkdsk	Check the status of a disk. **Scandisk** is preferred.
cls	Clear screen.
copy	Copy one or more files to specified location.
date	View or change the date.
defrag	Defragment a disk.
del, erase	Delete the specified files.
deltree	Delete a directory and all its contents, including other subdirectories.
dir	Display a directory.
diskcomp	Compare the contents of two floppy disks of the same size.
diskcopy	Copy all the contents of one floppy to another of the same size.
doskey	Run the doskey program.
dosshell	Start graphical interface to MS-DOS.
edit	Starts a full-screen editor.
exit	Exit the MS-DOS command interpreter.
fc	Compare two files.
fdisk	Configure hard disks for use with MS-DOS.
find	Search for a specified string in a file or files.
format	Format a disk for use with MS-DOS.
graphics	Allow DOS to print some graphic images.
help	Start the MS-DOS help system.
interlink	Connect two computers together via serial or parallel ports.

intersvr	Start the interlink server.
label	Change the volume label of a disk.
md, mkdir	Make a new directory.
mem	Display the amount of used memory and free memory.
memmaker	Optimize the computer memory.
mode	Configure system devices.
more	Display output one screen at a time.
move	Move files or rename directories.
msav	Scan for viruses.
msbackup	Back up and restore files.
mscdex	Provide access to a CD-rom.
msd	Display diagnostic information.
path	Display or change the search path.
print	Print a text file in the background.
prompt	Configure appearance of the DOS prompt.
qbasic	Start BASIC.
rd, rmdir	Remove a directory.
ren, rename	Rename a file.
replace	Replace files in destination directory with file of the same name in source directory.
scandisk	Disk analysis and repair tool.
smartdrv	Create a disk cache in extended memory.
sys	Create a bootable disk by copying hidden DOS system files.
time	Display or change the system time.
tree	Display a disk's directory structure.
type	Display the contents of a text file.
undelete	Attempt to recover a previously deleted file.
unformat	Attempt to unformat a previously formatted disk.
ver	Display MS-DOS version information.
verify	Toggle verify on or off.
vol	Display volume label and serial number.
xcopy	Copy one or more files and directories.

Windows 95 ® Basics (See Figure 6.12)

Mouse Techniques:

• Point	Roll the mouse to position the tip of the arrow on the screen over the desired icon.
• Click	Press and release the left mouse button.
• Right-click	Press and release the right mouse button.
• Double-click	Press and release the left mouse button quickly, twice in succession.
• Press	Press and hold the left mouse button.
• Drag	Press and hold the left mouse button while moving the mouse.

Menus: Menu items, which generally appear across the top of the window, may be selected with the mouse, or often with a keyboard shortcut. To use the shortcut, hold the control key down and press the listed letter key.

• Ctrl-N	New		• Ctrl-X	Cut
• Ctrl-O	Open		• Ctrl-C	Copy
• Ctrl-S	Save		• Ctrl-V	Paste
• Ctrl-P	Print		• Ctrl-F	Find

The Task Bar is located along the bottom of the Windows screen. It contains the **Start** button (at the left) and an icon for each active application (program.) Click on the **Start** button to open a menu that provides access to:

• Programs	All of the applications (such as data bases, word processors, and drawing programs) available through Windows.
• Documents	Rapid access to the most recently used files.
• Settings	Control Panel, Printer, and Task Bar settings.
• Find	Performs searches for files.
• Help	Starts *Windows Help*. (Not the same as the help utility provided with many programs.)
• Run	Prompts you to enter the name of a file to open. Files may also be run by double-clicking on the file icon, or by right-clicking on the file icon and clicking *Open*.
• Shut Down	Always click on this button before turning off the power to your computer.

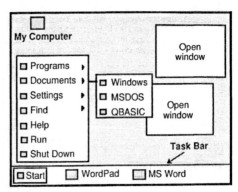

Figure 6.12 The basic parts of a Windows screen.

Parts of an open window. (Figure 6.13, next page). Open windows generally have the following features:

- Title Bar
This is the bar across the top of the window containing the program or document name. The color of the bar indicates the currently active window. Clicking anywhere in a window makes that window active. You can drag the active window around on the Windows screen by pressing and holding on the title bar.

- Minimize Button
The underlined button (third from the right, at the top) "minimizes" the window—that is, reduces it to an icon in the task bar. It can be maximized by double-clicking that icon.

- Restore Button
The box with the dark top portion toggles the window size between full-screen size and originally set size.

- Close
The "X" button at the top right corner closes the window.

- Menu Bar
Contains drop-down menu selections under a number of descriptive headings. **Save** and **Print** appear under **File**, for example. **Copy**, **Paste**, and **Find** appear under **Edit**.

215

- Scroll Bars Used to move the window over different parts of a large document. You can press the arrows to scroll slowly, drag the button to make large moves, or click on either side of the button to move one screen-full at a time.

- Screen You can change the size of a window by dragging an edge in or out. A double-headed arrow appears to show the direction.

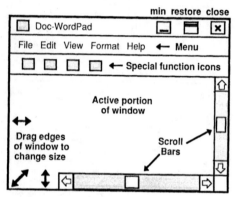

Figure 6.13 The basic parts of an active Window.

UNIX Instruction Summary

Following is a partial list of UNIX commands with their near-equivalents in MS-DOS.

UNIX	MS-DOS	Description
cat	type	Display contents of a file.
cd, chdir	cd, chdir	Changes directory.
chmod	attrib	Set file attributes.
cmp, diff	comp, fc	Compare files.
cp	copy, xcopy	Copy files.
date	date	Display or set the date.
date	time	Display or set the time.
echo	echo	Display a comment.
emacs, pico	edit	Full screen editor.
exit	exit	Exit from a shell.
grep	find	Search files for string.
jobs	(none)	List the processes by job.
kill	(none)	Kill a process or a job.
logout	(none)	End session, prevents access.
lp. lpr	print	Print a file.
ls -l	dir	List a directory of files.
mkdir	md, mkdir	Make a new directory.
more	file.txt>more	Display text file on screen.
mv	ren	Move (rename) a file.
passwd	(none)	Change your password.
ps	(none)	List the processes running.
pwd	cd	Display working directory.
rm	del, erase	Remove files.
rmdir, rm -r	rd, rmdir	Remove directories.
set PATH	path	Set the search path.
set, setenv	set	Set environmental variable.
sort	sort	Sort the contents of a file.

emacs Commands (UNIX Full Screen Editor)

Most terminals are equipped with special-purpose keys, such as **Home**, **End**, **Del**, etc. The functions of these should be obvious. The following is a partial list of Unix emacs commands.

Ctrl	undo	Ctrl-o	insert new line
Ctrl-a	beginning of line	Ctrl-p	previous line
Ctrl-b	back character	Ctrl-s	search forward
Ctrl-d	delete character	Ctrl-w	cut text to marker
Ctrl-e	end of line	Ctrl-y	yank (paste)
Ctrl-f	forward character	Ctrl-@	set marker in text
Ctrl-g	abort command	Ctrl-x Ctrl-c	exit emacs
Ctrl-h	help	Ctrl-x Ctrl-f	open file
Ctrl-t	tutorial	Ctrl-x Ctrl-s	save file
Ctrl-k	cut to end of line	Ctrl-x Ctrl-w	write file
Ctrl-n	next line		(save as..)

FORTRAN Statement Summary

CALL subroutine name (arguments)

CHARACTER name1∗len1, name2∗len2

CONTINUE

DATA name list / constant value list

DIMENSION name(size),name(row size,column size)

DO label, index name = initial, final, increment

END

FORMAT (field descriptions and line descriptions)

GOTO label

IF (logical expression) **THEN**   statements

IF (logical expression) **THEN**   statements
ELSE   statements
END IF

INTEGER list of names

PRINT (control list) output list

READ (control list) input list

REAL list of names

RETURN

STOP

SUBROUTINE subroutine name (arguments)

WRITE (control list) output list

Pascal Statement Summary

```
{comments}                    Comments in braces

x := 2                        Assignment of value

program progname;             Program heading

var x,y:real; ch1:char;       Variable declaration

procedure procname (x, y: real; n:integer);

function funcname (n:integer);

begin {progname}              General form of Pascal program,
statements                    procedure, or function
end; {progname}

read (x,y)                    Input and output
readln (a,b)                  statements. Suffix
write ('Result is', x)        ln causes jump to
writeln (x,y)                 a new line.

if x = 2 then                 If-then-else
    writeln ('Found it')      construct
else
    read(y);

for y := 0 to 9 do            Ten loops
    begin                     printing square
        x := sqrt(y);         roots.
    writeln (x)
    end;

while x > 0 do                Print squares of
    begin                     numbers until
        writeln (x*x)         numbers reach zero.
        x := x - 1;
    end;
```

```
repeat                          Increase bal by 8%
    bal := bal * 1.08;          for 20 iterations.
    iter := iter + 1
until inte = 20;


case y of
    0: writeln ('Y is zero');
    1,2,5,7,9: writeln ('Y is odd')

    writeln ('Y is even')
end;
```

Pascal Functions:

abs	sqr	cos	ln
round	sqrt	sin	odd
trunc	arctan	exp	

Reserved Words:

and	end	mod	repeat
array	file	nil	set
begin	for	not	string
case	forward	of	then
const	function	or	to
div	goto	packed	type
do	if	procedure	until
downto	in	program	var
else	label	record	while
			with

C-Language Summary

Control Statements

```
break;          Used to break out of a switch, while, do while,
                or for loop.
```

```
{
    declarations;       Compound statement block;
    statements;         Executes statements and hides
                        inside variables from
}                       outside processing.
```

`continue;`	Used with while, for, and do-while to skip to end of loop, then continue executing loop.

`do`
    `statement;`
`while (expression);`
	Execute statement repeatedly until expression evaluates to zero or false.

`for (ex1; ex2; ex3)` `statement;`	Evaluates expr1; repeatedly executes statement while ex2 is true; exits loop and evaluates ex3 when ex2 is false.
`goto label` `label: statement;`	Execution jumps to statement preceded by label. Keep label in same function as goto!

`if (expression)`
    `statement1;`
`else`
    `statement2;`
	Statement1 is executed if expression is true. Otherwise statement2 is executed.

`switch (expression) {`
  `case label1: statement1;`
           `break;`
  `case label2: statement2;`
           `break;`
  `- - - -`
  `default: statement3;`
       `break;`
`}`
	The expression is evaluated and compared to labels. If a match is found, the statement after the label is executed. Default case optional. Expression must be integer-valued.

`while (expression)` `statement;`	Statement executed repeatedly while expression is nonzero.
`return;`	Used in a function to return from that function to the statement following the function call.

Data Types

char	unsigned char	float	double	void
int	short int	long int	unsigned int	

Standard Functions in "C"

fopen	sscanf	log10	isalpha
fflush	putchar	pow	isalnum
fclose	puts	sqrt	isdigit
setbuf	putc	ceil	isxdigit
fseek	fputc	floor	iscntrl
ftell	fputs	strlen	isprint
feof	fwrite	strcpy	isgraph
ferror	printf	strcat	ispunct
clearerr	fprintf	strcmp	isspace
ungetc	sprintf	strncpy	atof
getchar	sin	strncmp	atoi
gets	cos	strchr	atol
getc	tan	strrchr	strtod
fgetc	asin	strspn	rand
fgets	acos	strcspn	srand
fread	atan	strstr	exit
scanf	exp	islower	abs
fscanf	log	isupper	

=	assignment	d	decimal	s	string
==	test if equal	i	integer	p	pointer
!=	test not equal	o	octal	f	floating point
>	test greater than	u	unsigned	e	exponential
<=	test equal or less	x	hex	g	e or f format
&	bitwise AND	!	not	c	single char
^	bitwise XOR	+	add	<<	left shift
I	bitwise OR	–	subtract	>>	right shift
&&	logical AND	*	multiply	++x	increment
II	logical OR	/	divide	––x	decrement
&x	address of x	%	modulus (remainder)		

Escape Characters

\b	backspace	\\	backslash
\f	formfeed	\"	double quotes
\n	newline	\'	single quote
\r	carriage return	\(CR)	line continuation
\t	horizontal tab	\nnn	octal character value
\w	vertical tab		

C++ Programming Language Summary

Comments: /* start comment   */ end comment
                // single-line comment

Reserved words:

asm	do	inline	sizeof
auto	double	int	static
break	else	long	struct
case	enum	new	switch
char	extern	operator	this
class	float	overload	union
const	for	public	unsigned
continue	friend	register	virtual
default	goto	return	void
delete	if	short	while

Constants. A sequence of digits is a decimal number unless it begins with one of the following: **0** (digit zero) = octal.
     0x or **0X** (digit zero and cap or lc eks) = hexadecimal.

Escape characters.

\a	beep	\\	backslash
\n	new line	\nnn	octal char value
\v	vertical tab	\f	form feed
\"	double quote	\t	horizontal tab
\b	backspace	\'	single quote
\r	carriage return	\0	null terminator

Scopes.   *Local*: A name declared in a block.
             File: A name declared outside any block or class.
             Program: A name declared externally.
             Class: The name of a class member is local to its class.

Storage classes.
   Automatic: Local to each invocation of a block and discarded upon exit from it.
   External: Visible throughout the entire program.
   Static: Retain values throughout entire program.

Data types.   int   char   float   long   double
                      const   unsigned int   unsigned char

Preprocessor.

#define identifier	#ifdef identifier
#endif	#include <filename>
#include "filename"	#else
#if expression	#ifndef identifier

Operators (decreasing precedence by group).

scope resolution	*class_name* :: *member*
global	:: *name*
select member	*object . member*
select member	*pointer -> member*
subscript	*pointer* [*expr*]
size of	sizeof [*expr*]
post increment	*x*++
pre increment	++*x*
post decrement	*x*- -
pre decrement	- -*x*
complement	~*x*
not	!*x*
unary minus	- *x*
unary plus	+*x*
address of	&*x*
de-reference	**x*
allocate memory	new *type*
de-allocate memory	delete *pointer*
de-allocate array	delete [] *pointer*
type convert	(type) *x*
member section	*object . * member_ pointer*
member section	*pointer ->* member_ pointer*

Unless otherwise indicated, the following operators take
the form of: *expr operator expr*

multiply, divide	* , /
modulo (remainder)	%
add, subtract	+ , −
less than, greater than	< , >
less than or equal	<=
greater than or equal	>=

equal, not equal	== , !=
bitwise AND	&
bitwise exclusive OR	^
bitwise inclusive OR	I
logical AND	&&
logical inclusive OR	II
conditional expression	*expr ? expr : expr*

assignment, modulo assign	= , %=
multiply assign	*=
divide assign	/=
add assign, subtr assign	+= , -=
left shift assign	<<=
right shift assign	>>=
AND assign	&=
inclusive OR assign	I=
exclusive OR assign	^=

I / O Operations.

cout << expression;	output
cout << setw(n);	set width of output field
cout << endl;	end line
cout << setprecision(n);	set decimal places
cout << setf(ios::left);	left justify field
cin >> variable;	input

Control Statement.

```
if (expr) statement;

if (expr)
     statement;
else
     statement;

while (expr) statement;
do
     statement;

while (expr);
```

```
for (expr; expr; expr)          for (i = 1; i<=3; i++)
     statement;                 executes statement 3 times
                                i = 1, i = 2, i = 3


switch (expr) {
     case constant:
          statement;
          break;
     default:
          statement;
          break;
}
```

Classes.
```
class class_name {
     public:
     // public data and methods go here
     ...
     private:
     // private data and methods here
     ...
};
```

Arrays.

Define:	int list [SIZE]
Initialize:	int list [] = {17, 01, 326, 3827}
Use:	list[i] ,

First element in array is a list[0]

QBASIC Programming Language Summary

Abbreviations Used.

[ ]	optional items	expr	expression
{ }	at least one required	num	number
I	OR separator	var	variable

Boolean Operators.

NOT	bit-wise complement	XOR	exclusive OR
AND	conjunction	EQV	equivalence
OR	inclusive OR	IMP	implication

Data Types.

$	string variable	%	integer
!	single precision	&	long integer
#	double precision		

INTEGER	16-bit signed integer variable
LONG	32-bit signed integer variable
SINGLE	32-bit floating-point variable
DOUBLE	64-bit floating-point variable
STRING * n %	string variable n% bytes long
STRING	variable-length string variable

ABS(num)	• Returns absolute value of num
ACCESS{READ I WRITE I READ WRITE}	
	• Specifies type of file access
ASC("A")	• Returns ASCII code of A
ATN(num)	• Returns arctangent of num (radians)
BEEP	• Beeps the speaker
BLOAD	• Load a file into memory
BSAVE	• Saves memory to a file
CALL	• Starts a procedure
CALL ABSOLUTE([ArgumentList,]Offset%)	
	• Pass control to machine language procedure
CDBL(num)	• Converts num to double precision
CHAIN"C:\TEMP\TEST.BAS"	
	• Transfer control to another BASIC program
CHDIR"C:\TEMP"	• Change directory
CHR$(65)	• Returns ASCII code of 65
CINT(num)	• Rounds num to an integer
CIRCLE [STEP] (x!,y!), radius! [, [color%] [, [start!]	
[, [end!] [, aspect!]]]]	• Draws a circle or ellipse

CLEAR [, , Stack&]	• Initializes variables & sets stack size
CLNG(num)	• Rounds num to a long
CLOSE [[#] FileNum% [, [#] FileNum%]...]	
	• Closes one or more open files
CLS [{0 I 1 I 2}]	• Clears screen
COLOR [Foreground%] [, [Background%] [,Border%]]	
	• Sets the screen colors
COM(n%) ON	• Enables event trapping on COMn%
COM(n%) OFF	• Disables event trapping on COMn%
COM(n%) STOP	• Suspends event trapping on COMn%
ON COM(n%) GOSUB label	• Directs events to subroutine
COMMON [SHARED] VariableList	
	• Declares global variables
CONST PI=3.14	• Defines PI as a constant
COS(angle)	• Returns cosine of angle in radians
CSNG(num)	• Converts num to single precision
CSRLIN	• Returns current row of the cursor
CVD(8-byte-numeric-string)	
	• Returns numeric equivalent of string
CVDMBF(8-byte-numeric-string)	
	• Returns IEEE-format equivalent of string
CVI(2-byte-numeric-string)	• Numeric equivalent of string
CVL(4-byte-numeric-string)	• Numeric equivalent of string
CVS(4-byte-numeric-string)	• Numeric equivalent of string
CVSMBF(4-byte-numeric-string)	
	• Returns IEEE-format equivalent of string
DATA 1 "Hello"	• Data for READ statements
DATE$	• Returns date in form mm-dd-yyyy
DATE$ = "01-02-98"	• Sets system date
DECLARE{FUNCTION I SUB}name [([ParameterList])]	
	• Declares a function or subroutine
DEFDBL, DEFINT, DEFSNG, DEFSTR	
	• Sets default data types
DIM A (3,5)	• Defines an array
DO WHILE n% < 5	• Do - while loop
n% = n% + 1	
LOOP	
DRAW	• Draws an object
ENVIRON "PATH = C:\DOS"	
	• Sets a DOS environmental variable
ENVIRON$ ('PATH')	• Returns a DOS environ variable
EOF (FileNum%)	• Returns nonzero at end of file

ERASE ArrayName	• Reinitializes array
ERDEV	• Returns the last error code
ERDEV$	• Returns name of device causing error
ERL	• Returns line number of last error
ERR	• Returns error code of last error
ERROR Expr%	• Simulates an error
EXIT	• Exits a DO, FOR, FUNCTION, SUB procedure or DEF FN function
EXP (Num)	• Returns e (2.718...) raised to Num power
FIELD [#] FileNum%, Field Width% AS StringVar$	
	• Allocates variable space in RAM file
FILEATTR (FileNum%, Attribute%)	
	• Returns information about file
FILES [path]	• Displays files in specified directory
FIX (Num)	• Truncates Num to integer

```
     FOR I% = 1 TO 10                    • For-Next loop
           PRINT I%
     NEXT I%
```

FRE (Num or String$)	• Returns amount of free memory
FREEFILE	• Returns next available file number

```
     FUNCTION name [ (ParameterList) ] [ STATIC ]
           [ statements ]
           name = expression          • Defines a function
           [ statements ]
     END FUNCTION
```

GET [#] FileNum% [, [Record&] [, variable]]	
	• Read from a file
GET [STEP] (x1! , y1!) - [STEP] (x2! , y2!) , ArrayName [(Index%)]	• Captures a graphic image
GOSUB	• Calls a subroutine
GOTO label	• Goes to the specified label in program
HEX$ (Num)	• Returns a hexadecimal string of Num

```
     IF condition1 THEN                  • If-then-else
           [ statements ]
     ELSEIF condition2 THEN
           [ statements ]
     ELSE
           [ statements ]
     ENDIF
```

INKEY$	• Reads a character from the keyboard
INP (Port%)	• Reads a byte from specified I/O port
INPUT [;] ["prompt" { ; I , }] Vars	• Reads from keyboard

INPUT$ (n [, [#] FileNum%]) • Reads n chars from a file
INSTR ([start%,] string1$, string2$) • Returns position of
string2 in string1 starting at start%
INT (Num) • Returns largest integer <= Num
IOCTL [#] FileNum%, String$
• Sends String$ to a device driver
IOCTL$ ([#] FileNum%) • Returns info from device driver
KEY Key%, String$ • Assigns String$ to function Key%
KILL FileName$ • Deletes FileName$ from disk
LCASE$(String$) • Converts String$ to lower case
LEFT$(String$, n%) • Returns leftmost n% characters
LEN (String$ or Var) • Returns size of String$ or Var
LINE [[STEP] (x1!, y1!)]-[STEP] (x2!, y2!) [, [Color%]
[, [B | BF] [, Style%]]] • Draws line or rectangle
LINE INPUT [;] [“prompt” ;] Var$
• Reads a line from the keyboard or file
LOC (FileNum%) • Returns current position in file
LOCATE [Row%] [, [Col%] [, [Cursor%] [, Start%
[, Stop%]]]] • Moves cursor to specified location
LOCK ... UNLOCK • Controls file access
LOF (FileNum%) • Returns length of file
LOG (Num) • Returns natural log of Num
LPOS (LPT%) • Returns the number of characters
sent to LPT% (1, 2 or 3)
LPRINT [expression] [{ ; | , }] • Prints to LPT1
LTRIM$ (String$) • Removes spaces to left of String$
MID$ (String$,Start% [, Length%])
• Returns Length% chars from String$
MKD$ (Double#) • Converts Double to String
MKDIR "C:\TEMP\TEST" • Creates a directory
MKDMBF$(Double#) • IEEE Double to Microsoft Binary
MKI$ (Integer%) • Converts Integer% to String
MKL$ (Long&) • Converts Long& to String
MKS$ (Single!) • Converts Single! to String
MKSMBF$ (Double#) • Converts IEEE Double# to Bin
Num1 MOD Num2 • Returns remainder of Num1/Num2
NAME OldName$ AS NewName$ • Renames file or dir
OCT$ (Num&) • Returns octal string of Num&
ON ERROR {GOTO Label I RESUME NEXT}
• Enables error trapping
ON KEY (n%) GOSUB Label
• Specifies location of key-trapping subroutine

231

```
ON PLAY (QueueLimit%) GOSUB line
                    • Specify subroutine when buffer limit reached
ON STRIG (n%) GOSUB Label
                    • Specify subroutine for joystick event trapping
ON TIMER (n%) GOSUB Label
                    • Specify subroutine for timer event trapping
ON n% GOSUB Label1, Label2
                    • Do subroutine based on value of n%
ON n% GOTO Label1, Label2
                    • Go to Label based on value of n%
OPEN file$ [ FOR Mode ] [ ACCESS access ] [lock ]
    AS [ # ] fileNum% [LEN = reclen% ]
    file$   Name of file or device
    mode    APPEND, BINARY, INPUT, OUTPUT, RANDOM
    access  With networks: READ, WRITE, or READ WRITE
    lock    With networks: SHARED, LOCK READ,
                    LOCK WRITE, LOCK READ WRITE
    fileNum% 1 through 255
    reclen% Record length (default 128 bytes)
OPEN "COMn: opt1 opt2" [ FOR Mode ] AS [ # ]
    file% [LEN=reclen% ]           • Opens COM port
OPTION BASE { 0 | 1 }       • Sets lower bounds for arrays
OUT port%, data%            • Sends a byte to an I/O port
PAINT [ STEP ] (x!,y!) [ , [ {color% | tile$ } ]
    [ , [BorderColor%] [ ,BackColor$] ] ]  • Fill graphics area
PALETTE [attr%, color& ]            • Sets color palette
PCOPY source%,destination%
                    • Copy source% video memory to destination%
PEEK (address)              • Returns a byte from memory
PEN ON, PEN OFF, PEN STOP           • Light pen trapping
PLAY NoteString$                    • Play musical notes
PLAY (n)   • Returns number of notes in music queue
PLAY ON, PLAY OFF, PLAY STOP,       • Controls play
   ON PLAY (queue%) GOSUB label          event trapping
PMAP (viewport#, n%)    • Window coordinate to viewport
POKE address, byte%        • Write byte to memory address
POS (0)            • Returns current column of cursor
PRESET or PSET [STEP] (x!y!) [ , color%]  • Draw a point
PRINT [#File%,] [Expr] [ { ; | , } ]   • Print to screen or file
PRINT [#File%,] USING format$; Expr [ { ; | , } ]   • Print
PUT [ # ] File% [ , [record&] [ , Var ] ]   • Write to a file
PUT [STEP] (x1!, y1!), ArrayName [ (index%) ] [ , action]
                    • Display an image
```

232

```
RANDOMIZE [seed%]        • Init random-number generator
READ VarList             • Reads values from DATA statement
REDIM [SHARED] Var(subscripts) [AS type]
                         • Resize a dynamic array
REM                      • Makes remainder of line a comment
RESET                    • Closes all open files
RESTORE [label]          • Re-read data from DATA statement
RESUME [ {label | NEXT} ] • Resume program from error
RETURN [label]           • Return from subroutine
RIGHT$ (String$, n%)
                         • Returns rightmost n% characters from String$
RMDIR path$              • Removes the specified directory
RND                      • Returns a random number between 0 and 1
RSET StringVar$ = string$ • Move data to a RAM buffer
RTRIM$ (String$)         • Remove spaces from right of String$
RUN                      • Runs the specified program
SCREEN mode% [ , [ColorOnOff%] [ , [ActivePage%]
  [ ,VisualPage%] ] ]    • Sets screen mode
SCREEN (row%,column% [ , ColorFlag%] • Returns color
SEEK (FileNum%)          • Returns file position
SEEK [ # ] FileNum%, position& • Sets file position
    SELECT CASE TestExpr
        CASE expr1
            [statements]
        CASE expr2
            [statements]
        CASE ELSE
            [statements]
    END SELECT
SGN (Num)                • Returns 1 if > 0, 0 if = 0, –1 if < 0
SHARED variable [ ( ) ] [AS type]
    • Makes module-level variables accessible to procedures
SHELL [command$]         • Runs DOS command or batch file
SIN (angle)              • Returns sine of angle in radians
SLEEP [seconds&]         • Suspends program
SOUND frequency, duration • Makes a sound
SPACE$ (n%)              • Returns a string of n% spaces
SPC (n%)                 • Skips n% spaces in Print or Lprint
SQR (Num)                • Returns square root of Num
STATIC var [ ( ) ] [AS type] • Makes var local & preserves
STICK (n%)               • Returns coordinates of a joystick
STOP                     • Halts a program
```

233

STR$ (Num)	• Returns string representation of Num
STRING (n%) [ON, OFF, STOP]	• Joystick event trapping
STRING$ (length%, {ascii% l StringExpr$})	
	• Returns a string of repeating characters
SUB name [parameters)] [STATIC]	
statements	• Defines a subroutine
END SUB	
SWAP var1, var2	• Swaps two variables
SYSTEM	• Closes all files, returns to operating system
TAB (column%)	• Moves cursor
TAN (angle)	• Returns tangent of angle in radians
TIME$	• Returns current time
TIME$ = hh:mm:ss$	• Sets current time
TIMER [ON, OFF, STOP]	• Timer event trapping
TRON, TROFF	• Enable, disable tracing
TYPE usertype	• Define date types
name AS TypeName	
. . .	
END TYPE	
UBOUND (array) [,dimension%])	
	• Returns upper bound for array dimension
UCASE$ (String$)	• Converts String$ to upper case
UNLOCK [#] file% [, {record& l [start&] TO end&}]	
	• Unlocks specified file
VAL (String$)	• Converts String$ to a number
VARPTR (Var)	• Returns the offset address of Var
VARPTR$ (Commands$$)	• Returns Command$ adr as string
VARSEG (Var)	• Returns segment address of variable
VIEW [[SCREEN] (x1!, y1!)-(x2!, y2!)	
[, [color%] [,border%]]]	• Defines a viewport
VIEW PRINT [TopRow% TO BottomRow%]	
	• Sets boundaries of text display
WAIT PortNum%, AND-Expr% [, XOR-Expr%]	
	• Waits for specified bit pattern on PortNum
WHILE condition	• Executes statements
statements	while condition
WEND	is true
WIDTH [cols%] [, rows%]	• Sets screen columns and rows
WIDTH [#FileNum% l device$}, col%	• Sets output width
WIDTH LPRINT col%	• Sets printer output width
WINDOW [[SCREEN] (x1!, y1!) - (x2!, y2!)]	
	• Sets graphics window
WRITE [[#] FileNum%,] ExprList	• Writes to screen

7

Tests and Procedures

7.1 SOLDERING AND REWORKING

Soldering pencils. For working on circuit boards containing ICs or small semiconductors, use a temperature-regulated soldering pencil. The tip temperature of these pencils will not soar in the cradle, so the potential for damage to the circuit board is minimized. These pencils use iron-plated tips that maintain a smooth, corrosion-free surface if kept clean by frequent wiping on a wet sponge. These tips should never be filed or sanded.

If an unregulated pencil *must* be used on microelectronic circuits, select one with a power rating of 20 W or less. A 400-PIV diode in series with the line cord of a 30-W pencil will reduce the power to 15 W. Use unplated copper tips, and file them as necessary to keep them smooth and of the proper shape. After filing, wrap solder around the tip and turn the pencil on to "tin" the tip with a coating of solder as it heats up. Always wipe off excess solder on a wet sponge, as molten solder will dissolve the copper tip.

Solder. Eutectic (yoo-TEK-tik) solder has the lowest melting point (183 °C), and passes from liquid to solid without going through a pasty phase. This alloy is 63/37, or 63% tin, 37% lead, and is the most desirable. The alloy 60/40 melts at 190 °C and is also acceptable. Alloys 50/50 and 40/60 melt above 200 °C and should be avoided in electronic work.

Liquid solder absorbs oxygen from the air if it is overheated or held too long in the liquid state. Oxidized solder appears dull and grainy, and leaves jagged peaks when the pencil tip is pulled away. Such solder must be removed and fresh solder applied.

Flux. Metals cannot generally be bonded with solder and heat alone because an oxide forms on the surface of the metal which will not accept the solder. Most solder for electronics use has a core of rosin *flux* to clean off the oxide and permit bonding. On some special jobs, flux is added to the work separately, but in such cases the flux manufacturer's recommendations must be followed closely, since non-rosin fluxes are generally corrosive and often electrically conductive. Rosin is not conductive, but it should be cleaned off with alcohol or flux solvent after soldering is completed.

How to solder. *1*. The work must be as clean as possible to make the cleaning action of the flux effective. *2*. The work must be hot enough to melt the solder by itself. *3*. Complete the soldering operation in five seconds or less to avoid oxidizing the solder.

Use a wire brush, steel wool, or knife edge to scrape corroded parts clean. On stubborn jobs, tin each part with a coating of solder before joining them.

Heat the work with the pencil point from one side and apply the solder to the other side to assure that the work is hot enough to melt the solder. You may use a small dab of solder on the pencil point to serve as a heat bridge to the work. Press the point firmly to the work to get full heat transfer.

A good solder joint on a pc board looks like a chocolate kiss—smooth and concave. A bad joint looks like an onion dome—bulging and not adhered down to the board.

Desoldering. Never pull on component leads while heating their connections at the pc board. The glue that holds the tracks to the board is weakened by heating, and you will pull the tracks off the board. Use a solder sucker or solder-absorbing braid to remove all the solder from the joints; then pull the component out when the joints have cooled. Quickly clamp the solder sucker over the hole and actuate it to get the joint completely clear of solder. If the component costs less than a dollar, it is wiser to clip all its leads and remove them one by one; you can afford to lose the dollar, but not the whole circuit board. Wooden toothpicks or sewing needles can be used to clear the holes while heating prior to insertion of the new component.

7.2 COMPONENT TESTS

Low-value resistances, such as a printed-circuit run or a coil winding, can be measured more accurately by passing a relatively large measured current through the component and measuring the voltage drop across it. Use a current-limiting resistor in series with the supply and the component to keep the test current from becoming excessive. Calculate

$$R_{series} = \frac{V_{SUPPLY}}{I_{TEST}} \qquad P_{R\,(series)} = I_{TEST} \times V_{SUPPLY}$$

Clip the voltmeter leads directly to the component when measuring its voltage (V_x) to avoid measuring the IR drop on the current-carrying leads. An example is given with Figure 7.1, below.

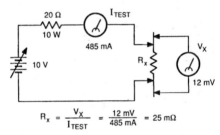

$$R_x = \frac{V_x}{I_{TEST}} = \frac{12\ mV}{485\ mA} = 25\ m\Omega$$

Figure 7.1 Measuring milliohm-value resistances.

Delicate resistances, such as meter coils, may be damaged simply by attempting to measure them with an ohmmeter. Instead, pass a small measured current through the resistance with a dc supply and high-value series resistance. Measure V_x across the component and calculate $R = V / I$. Figure 7.2, on the following page, provides an example.

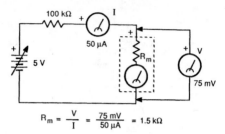

$$R_m = \frac{V}{I} = \frac{75\ mV}{50\ \mu A} = 1.5\ k\Omega$$

Figure 7.2 Never measure the resistance of a delicate meter coil directly. Calculate it from measured voltage and current.

High-value resistances beyond the range of your ohmmeter may be conveniently measured by using a DVM on the dc voltage scales to form a voltage divider with the unknown resistance, as shown in Figure 7.3, below. The fraction of the supply voltage that appears across the DVM is determined by the DVM's resistance and that of the unknown resistor. The 0.1-μF bypass capacitor shown is sometimes required to prevent the high-impedance voltmeter from picking up too much noise.

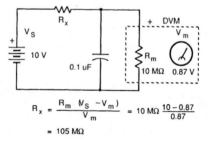

$$R_x = \frac{R_m\ (V_S - V_m)}{V_m} = 10\ M\Omega\ \frac{10 - 0.87}{0.87}$$

$$= 105\ M\Omega$$

Figure 7.3 Measuring resistance values higher than the range of the DVM ohmmeter.

Resistance values higher than the range of the DVM can be determined without the dc source required in Figure 7.3 by the setup and calculation shown in Figure 7.4, below. The trick is to shunt the unknown resistance with a resistance of a known value that will bring the parallel combination within the range of the ohmmeter.

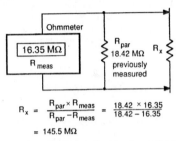

$$R_x = \frac{R_{par} \times R_{meas}}{R_{par} - R_{meas}} = \frac{18.42 \times 16.35}{18.42 - 16.35}$$

$$= 145.5 \ M\Omega$$

Figure 7.4 Another method for measuring high-value resistances.

Battery testing should be done under load, as shown in Figure 7.5, below. Where the rated or required output current is known, the value of R_L may be calculated from V_S / I. The formula given with the figure is a rule of thumb based on battery size (volume). Don't forget to calculate the power-dissipation requirement of R_L.

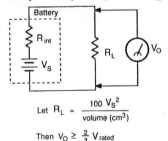

$$Let \ R_L = \frac{100 \ V_S^2}{volume \ (cm^3)}$$

$$Then \ V_O \geq \frac{2}{3} V_{rated}$$

Figure 7.5 Testing a battery under load.

Nonpolar capacitors from 10 μF to 100 pF can be measured to an accuracy of about 3% with the test setup of Figure 7.6, below. The generator output is preset to 10.0 V on the oscilloscope or VOM, and the frequency is varied to produce 1.00 V across the 100-Ω 1% resistor. V_g is then rechecked and adjusted if necessary, and C is calculated from the formula given. Do not use an ordinary DVM to measure the ac voltages, as DVMs are generally extremely inaccurate in measuring voltages at frequencies above a few kilohertz. VOMs typically can be trusted to a few hundred kilohertz, and most modern oscilloscopes to several megahertz.

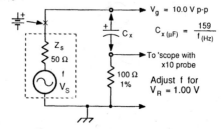

Figure 7.6 Capacitor measurement.

Polar capacitors can be measured using the same circuit with the addition of a 6- or 9-V battery as shown to bias the capacitor properly. For values above 10 μF, the 100-Ω resistor may be replaced with 10.0 Ω, in which case the formula becomes

$$C_{(\mu F)} = \frac{1590}{f_{(Hz)}}$$

Inductor values from 10 H to 100 μH can be measured using the test circuit of Figure 7.7 (next page) and the procedure described above. If the Q of L_x becomes less than about 5, the accuracy of the determination of L_x will be seriously impaired. The range can be lowered to 10 μH if R is changed to 10.0 Ω, in which case the formula becomes

$$L_{(H)} = \frac{15.9}{f_{(Hz)}}$$

240

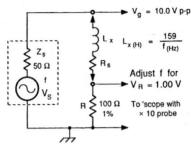

Figure 7.7 Inductor measurement.

Inductor Q from 1 to about 200 can be measured with the circuit of Figure 7.8, below. The inductor value L_x is first measured as in Figure 7.7. Since Q varies with frequency, the test frequency f must be selected to be near the intended operation frequency. Now calculate a capacitance C that will resonate with L_x at frequency f. Choose an available high-Q capacitor C_s near to the calculated value and vary f until V_C peaks at f_r. Finally, measure V_g. The ratio V_C/V_g equals Q, since X_L and X_C cancel at f_r and V_g equals $V_{R(s)}$. The 2.2-Ω resistor prevents V_g from soaring and dipping wildly as the load on the generator varies near f_r. If Q is very high and the generator waveform contains harmonics, V_g may become very distorted, invalidating the measurement.

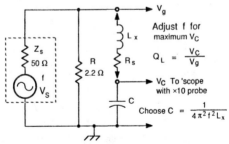

Figure 7.8 Measuring inductor Q.

Saturation current for ferrite-core inductors can be measured with the setup of Figure 7.9 below. The transformer secondary must have a current rating higher than the inductor under test. V_S is increased until the oscilloscope, in the ac coupled mode, shows a marked decrease in the ac signal across the inductor. A 60-Hz test frequency may be used, but 200 Hz is easier to view on the 'scope. The formulas given convert I_{sat} to $V_{sat(pk)}$ under sine-wave excitation at frequency f, and to volt-microsecond constant for pulse transformers.

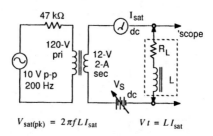

$$V_{sat(pk)} = 2\pi f L I_{sat} \qquad V t = L I_{sat}$$

Figure 7.9 Measuring inductor saturation current.

Transformer coefficient of coupling k, and turns ratio n, can be determined by the measurements and formulas shown in Figure 7.10, below. V_1 and V_2 are measured with the generator connected to one winding, and then V_3 and V_4 are measured with the generator across the other winding. If k is near unity, the turns ratio becomes simply V_2/V_1.

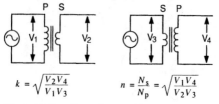

$$k = \sqrt{\frac{V_2 V_4}{V_1 V_3}} \qquad n = \frac{N_s}{N_p} = \sqrt{\frac{V_1 V_4}{V_2 V_3}}$$

Figure 7.10 Measuring transformer coefficient of coupling and turns ratio.

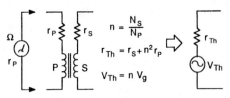

$$n = \frac{N_S}{N_P}$$

$$r_{Th} = r_S + n^2 r_P$$

$$V_{Th} = n V_g$$

Figure 7.11 Obtaining a Thévenin equivalent of audio and power transformers by resistance measurements.

Most audio and power transformers can be represented by a Thévenin equivalent circuit obtained from simple resistance measurements and the formulas given with Figure 7.11, above. For audio transformers above 50 W and power transformers above 500 W, and for any transformer operating above the audio-frequency range, skin effect is likely to invalidate resistance readings taken with a dc ohmmeter, and it will be necessary to load the transformer and take a voltage-drop measurement, as shown in Figure 7.12, below, to obtain a Thévenin equivalent circuit. Choose R_L experimentally to cause about 10% voltage drop at the secondary, and be *sure* to calculate its power requirement.

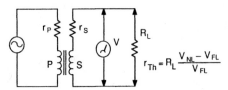

$$r_{Th} = R_L \frac{V_{NL} - V_{FL}}{V_{FL}}$$

Figure 7.12 Obtaining the Thévenin resistance of high-power and high-frequency transformers by measuring voltage drop under load.

7.3 SEMICONDUCTOR TESTS

Diodes may be tested by simply checking for a low forward and high reverse resistance on a VOM or DVM. A few checks on known-good diodes will permit you to mark your VOM leads for determination of anode and cathode on unknown types. Experience will also allow you to distinguish among silicon, germanium, and Schottky diodes by the progressively lower on-state resistance reading. Diodes may be matched for identical on-state resistance by this technique. Try to settle on one scale of your ohmmeter (1-kΩ or ×1 kΩ) for these tests to avoid confusion. The "Low R" range, available on some DVMs, does not refer to low resistance values; it applies a low *test voltage*, below 0.5 V, so silicon diodes will read as very high resistances either way.

Diode recovery times t_{rr} may be compared among power diodes with the simple test setup of Figure 7.13, below. Fast diodes are required for inverters, switching regulators, and flyback power supplies.

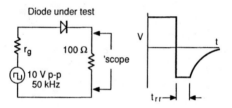

Figure 7.13 Measuring diode recovery time.

Zener or avalanche diode voltage may be checked with the test setup of Figure 7.14, below. The source voltage should be from 1.5 to 3 times V_Z.

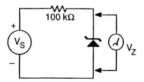

Figure 7.14 A simple setup for measuring zener diode regulator voltage.

244

Peak Inverse Voltage (PIV), also called Peak Reverse Voltage (PRV), of rectifier diodes can be measured nondestructively with the setup of Figure 7.15, below, if a variable high-voltage dc supply is available. V_S is increased until the microammeter just begins to indicate current. V_S then is approximately equal to diode PIV. Be careful not to allow a current of more than a few microamps or the diode may be damaged. Also, for protection against high voltages, be sure to turn off the supply and short its output with an insulated jumper cable before making any changes to the circuit. Take care that no one touches the circuit while the power is on.

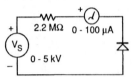

Figure 7.15 Measuring diode PIV. Observe safety precautions when working with high voltages.

Bipolar transistors may be tested as two diode junctions (base-emitter and base-collector) as shown in Figure 7.16. To identify the leads, find a pair that measure high resistance both ways; the remaining lead is the base. The base must show high/low resistance to each of the other leads as the ohmmeter is reversed, or the transistor is bad (or it is not a bipolar transistor at all).

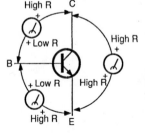

Figure 7.16

A bipolar transistor shows high/low resistance from B to E and B to C. PNP types show low/high resistances. Both types show high / high resistance between C and E.

Transistor beta can be estimated and compared among units by the simple test illustrated in Figure 7.17, below. The ohmmeter provides a voltage source and a current meter for the collector-emitter main current path. With the base left open, the collector-emitter will conduct only the leakage current I_{CEO}, which is generally observable only in older germanium transistors. When current is fed from collector to base through a 100-kΩ resistor (or by simply grasping the collector and base leads with the fingers and using skin resistance) the transistor will turn on, showing a "resistance" reading much lower than the 100 kΩ resistor.

This test can be used to distinguish collector from emitter where the case does not make the pinout clear, since silicon planar transistors will show very low current gain if the collector and emitter are reversed.

Power transistors may test more successfully on a lower range of the ohmmeter because of their higher leakage and lower beta at low currents. Use a 1-kΩ collector-base resistor in this case. For PNP transistors, reverse the probes to place the negative test voltage at the collector.

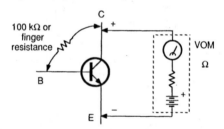

Figure 7.17 Ohmmeter transistor beta check.

Field-effect transistors can be checked for certain faults with an ohmmeter. The gate-source should show a diode junction for JFETs and open both ways for MOS types, but in no case low resistance both ways. The drain-source with the gate tied to the source should show several hundred ohms for depletion types and nearly infinite for enhancement types, but in no case near zero.

The test circuit of Figure 7.18 may be used to test almost any small-signal FET for transconductance y_{fs}. Enhancement types will show control of I_D for positive V_{GS} whereas depletion types will require negative V_{GS} to control I_D. For P-channel FETs, reverse both polarities.

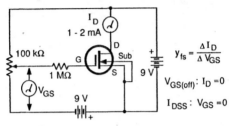

$$y_{fs} = \frac{\Delta I_D}{\Delta V_{GS}}$$

$V_{GS(off)}: I_D = 0$

$I_{DSS}: V_{GS} = 0$

Figure 7.18 A general test circuit for FETs.

Unijunction transistors should show a diode junction from E to B_1 and from E to B_2, and several hundred ohms from B_1 to B_2 with either polarity. A functional check can be made by putting the device in the simple test circuit of Figure 7.19.

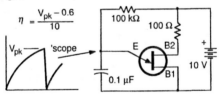

$$\eta = \frac{V_{pk} - 0.6}{10}$$

Figure 7.19 A unijunction transistor test circuit.

SCRs and triacs may be tested with the circuit of Figure 7.20. The lamp should go on when S_1 is pressed and stay on until V_S is removed. Triacs should perform equally with V_S reversed. For low-power SCRs and triacs a VOM on the R×1 range connected from anode to cathode will show high resistance until the anode is shorted briefly to the gate, whereupon the resistance will go low and remain there until the VOM is disconnected from the anode.

247

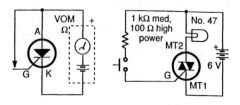

Figure 7.20 Tests for low-power and higher-power SCRs and triacs.

7.4 AMPLIFIER TESTS

Input impedance may be determined with the test setup of Figure 7.21, below, if it is purely resistive. R_v is first set to zero and V_o is noted, taking care to stay well below the level of distortion. R_v is then adjusted until V_o drops to one-half of its original value. R_v is then removed and measured. Since one half of V_s appeared across R_v and the other across Z_{in}, the two resistances R_v and Z_{in} must be equal. If Z_{in} is not at least 50 times greater than Z_s of the generator, it will be necessary to shunt the generator with a resistance about $\frac{1}{50} Z_{in}$ to prevent changes in generator output voltage as R_v is varied. We measure V_o rather than V_{in} because it is larger, and because it keeps the impedance of the 'scope from shunting a high-value Z_{in} and invalidating the measurement.

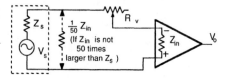

1) Set $R_v = 0$ and note V_o.
2) Adjust R_v so V_o drops to one half.
3) R_v now equals Z_{in}.

Figure 7.21 Measuring amplifier input impedance.

Output impedance is measured as shown in Figure 7.22. Measure V_o with no load connected; then add a variable load and adjust until V_o drops to one half of its original value. Since one half of V_a (in the figure) appears across Z_o and the other appears across R_v, the two resistances must be equal, and R_v can be removed and measured on an ohmmeter to determine Z_o. If distortion of the output waveform becomes evident as R_v is reduced, lower V_{in}.

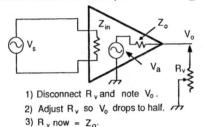

1) Disconnect R_v and note V_o.
2) Adjust R_v so V_o drops to half.
3) R_v now = Z_o.

Figure 7.22 Measuring amplifier output impedance.

Harmonic distortion down to about 1% can be measured with the test setup of Figure 7.23. Commercial instruments based on the same principle can measure lower distortion figures. With the signal generator at the input of the T-notch filter, the output should drop to 0.5% or less of V_s. Connect the amplifier to be tested between V_s and the notch filter, and adjust V_s for the rated V_o to the load. Finally, measure V_{dist}, preferably with a true-rms voltmeter. The notch-filter values shown set the test frequency to 400 Hz.

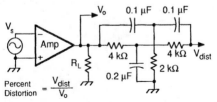

$$\text{Percent Distortion} = \frac{V_{dist}}{V_o}$$

Figure 7.23 Measuring harmonic distortion.

Common-Mode Rejection Ratio (CMRR) is measured by comparing output responses for the two input connections shown in Figure 7.24. If A_v is very high or CMRR is very low, V_1 may become distorted or it may become impossible to see V_2 on the oscilloscope. In this case, make $V_{in\ (com)}$ 100 times $V_{in\ (diff)}$ and multiply CMRR by 100. Common-mode rejection ratio is generally specified in decibels, rather than as a voltage factor, so the conversion equations are repeated here.

$$CMRR_{db} = 20 \log \frac{V_2}{V_1} \qquad \frac{V_2}{V_1} = 10^{(CMRR/20)}$$

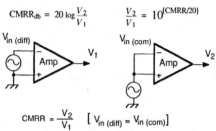

$$CMRR = \frac{V_2}{V_1} \quad [\ V_{in\ (diff)} = V_{in\ (com)}\ ]$$

Figure 7.24 Measuring common-mode rejection ratio.

Common-mode voltage limits may be determined with the setup of Figure 7.25. An ac signal is applied differentially through a transformer, and is adjusted to produce an amplifier output not greater than 5% of maximum. A dc voltage is then applied in common-mode, and increased until the ac portion of the output falls below the gain or distortion specifications. The dc input value is then at the common-mode voltage limit. The test is repeated with the dc supply reversed to obtain the negative common-mode voltage limit.

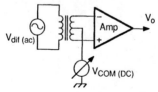

Figure 7.25 Measuring common-mode voltage limits.

7.5 MISCELLANEOUS TESTS AND TIPS

Hot chassis? A voltmeter test from a suspected chassis to earth ground may be misleading because stray capacitance normally couples relatively harmless currents to an ungrounded chassis. Connect a 47-kΩ, $1/2$-W resistor across the voltmeter and then take the measurement. Leakage currents should produce no more than 10 V across this load.

Percentage of modulation (AM). An RF-vs.-time or RF-vs.-AF display can be used to produce one of the patterns of Figure 7.26, from which percent modulation can be calculated. If the vertical bandwidth of the oscilloscope is less than the RF frequency, the RF will have to be coupled directly to the CRT deflection plates.

$$\text{Percent Modulation} = \frac{V_{pk} - V_{val}}{V_{pk} + V_{val}} \times 100\%$$

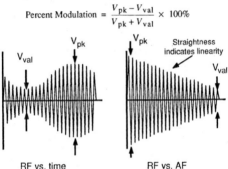

RF vs. time   RF vs. AF

Figure 7.26 Checking percent AM modulation.

A lissajous figure (say LEE sah jhoo) is obtained when two sine-wave signals are applied to the vertical and horizontal deflection systems of an oscilloscope. If there is an integer-ratio frequency relationship between the two signals, the figure will appear stable and will consist of one or a number of loops. The number of loops in one row counted horizontally gives the vertical frequency-ratio number, and the number of loops in one column vertically gives the horizontal frequency-ratio number.

In the simple case where the vertical and horizontal frequencies are equal, the figure will be an ellipse, and the phase difference between the two signals can be determined using the equation given with Figure 7.27, below.

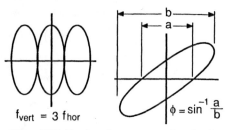

$$f_{vert} = 3\,f_{hor}$$

$$\phi = \sin^{-1}\frac{a}{b}$$

Figure 7.27 Lissajous figures compare two signals.

Risetime t_r is defined as the time it takes a waveform to go from 10% to 90% of its full value. Falltime is the time from 90% to 10%. For amplifiers whose square-wave response is essentially free from overshoot and ringing, the high-frequency –3-dB point and risetime are related as follows:

$$f_{high} = \frac{0.35}{t_r} \qquad t_r = \frac{0.35}{f_{high}}$$

Line noise? If a noise signal is encountered when troubleshooting with an oscilloscope, switch the trigger source to *line*. If the display holds still, the source of the noise is the ac line.

Estimating dc supply current without unsoldering anything is easy. Just measure the peak-to-peak ripple voltage and time between ripple peaks across the first filter capacitor and calculate

$$I_L = \frac{C_{filt}\,V_{rip\,(p\text{-}p)}}{t}$$

Is the zener conducting? An avalanche diode (7 V and above) will have about 2 to 20 mV of white noise across it if it is conducting. If the noise doesn't show up on a sensitive 'scope, it's a good bet that the zener isn't conducting.

Meter loading? If a VOM is loading a circuit significantly, the voltage reading will change with changing ranges (2.0 V on the 2.5-V range and 3.0 V on the 10-V range, for example). A DVM will not show this effect because its resistance is the same on all ranges. However, if two meters are available and the reading on the first changes significantly when the second is connected across it, there is a meter-loading problem.

Tape should never be regarded as a permanent fixture. A taped joint can be made to hold longer by placing a "flag" of tape with the two sticky surfaces mating around the original wrap, but even this will come loose with time. Use heat-shrink tubing for permanence.

Printed-circuit tracing is easier if you shine a light through the board from the bottom while viewing it from the top.

Self-oscillation at frequencies above 100 MHz may not show up on your oscilloscope, but if moving your hand near the wiring or touching the chassis causes changes in the instrument's operation, self-oscillation is likely to be the problem.

Metal transistor cases are usually connected to the transistor collector, and provide a convenient test point.

A soldering gun that won't heat up probably just needs the tip bolts tightened.

Cheap panel meters sometimes have no zero adjust. They can be rezeroed by holding the body of a soldering gun near and turning it on and off. Several tries may be needed to get the residual magnetic field just right.

A microammeter pointer will bounce around when the meter case is rotated, but it will hold quite steady if the leads are shorted first. You can tell if the meter coil is burned out in this way.

Plastic-face panel meters sometimes give false readings or act "sticky" because of static charge on the face. Simply moisten the plastic face.

Variable capacitor plates are delicate and easily bent. If you listen closely as you rotate them, you can hear a scraping if they are shorting together.

Equalizing resistors. Electrolytic capacitors are available with voltage ratings only up to about 500 V. To handle higher voltages, several capacitors may be placed in series. However, to equalize the dc voltage drops, the capacitors must be shunted with equal-value resistors. Values of 270 kΩ or 470 kΩ are typical for 50-μF, 450-V electrolytics. Remember to calculate the power dissipation of these equalizing resistors, and the reduced value of the series capacitance.

Series diodes. Diodes are sometimes placed in series to obtain a higher PIV rating. Equalizing resistors are not required, since reaching a diode's avalanche point is not in itself destructive; the other diodes will simply take the reverse voltage if one reaches avalanche. Equalizing *capacitors* may be required, however, to prevent one "faster" diode from taking all the PIV (and carrying enormous current) while all the other "slower" diodes are still in conduction. This is seldom necessary at 60 Hz, but is helpful in high-frequency or square-wave-driven rectifiers.

Diodes rated above 1 kV PIV often consist of several diodes in series in a single case. A VOM or DVM applying a test voltage of 1.5 V will not turn such a diode on in either direction. Some VOMs apply a sufficient test voltage on their highest resistance range.

Potentiometers tell a lot by the way they feel as they are turned. A bumpy feel indicates a wire-wound type. A single bump indicates a broken wire or carbon track. Irregular bumps indicate a burned carbon track. If noise diminishes when you pull on the knob, wiper tension may be weak or the wiper arm may need cleaning.

Low-value resistance. When measuring resistances below 10 Ω, short the ohmmeter leads first and check for a zero reading. If zero cannot be obtained, subtract this "lead resistance" from your reading.

Measuring over-range resistors. Many DMMs have an upper limit of 20 MΩ. Higher values can be measured by paralleling them with a previously measured resistor just below the range limit. For example, a nominal 18-MΩ resistor measures 18.34 MΩ. In parallel with R_x, the measurement is 10.15 MΩ.

$$\frac{1}{R_x} = \frac{1}{R_{par}} - \frac{1}{R_1} = \frac{1}{10.15} - \frac{1}{18.34} \; ; \; R_x = 22.7 \text{ MΩ}$$

High Ohms. Some DMMs have "High Ohms" and "Low Ohms" ranges. The *high* and *low* refer to the test voltages used, not to the resistance values being measured. On *Low Ohms* the applied voltage should be below 0.5 V, so silicon diode junctions will appear as open circuits. On *High Ohms* diode junctions will be turned on and register a low "resistance." This is a useful troubleshooting tool.

DVM frequency limits. Many DVMs lose accuracy or become completely useless at frequencies above a few kilohertz. It's best to use a 'scope to verify the frequency and waveshape of ac voltages above 60 Hz.

Never leave a VOM on the ohms range. It could discharge the battery, and if left for an extended period, the battery could leak corrosive paste all over the circuitry inside.

Loudspeaker cones can be repaired with a piece of tissue paper soaked in nail polish.

Clear PC-board holes with a wooden toothpick while desoldering.

Pencil lead is conductive. Never mark a circuit board with a pencil.

Intermittent problems often disappear when the instrument is taken out of the cabinet because it stays cooler with the air circulating freely around it. Use a 100-W hooded lamp to heat the suspected area of the instrument and a can of spray coolant to cool individual components until the problem is located.

Removing ICs. If you have to remove an IC from a circuit board, and its cost is less than $2.00, it will probably be better to clip all the pins and remove them one by one. Wrestling the whole thing out intact poses a great danger of damage to the board. If you have to scrap that you'll lose a lot more than $2.00!

Stripping wire. If you must strip wire with a knife blade, roll the insulation over the blade to avoid nicking and thus weakening the wire. Never use a sawing motion.

Toggle and pushbutton switch contacts bounce (typically five or ten times) for a few milliseconds on actuation. Rotary switches generally do not bounce. Normal relay contacts bounce, but reed relay contacts do not.

Normal fuses will blow on 200% overload in about 2 seconds. They have a thin wire or strap inside the glass tube. Slow-blowing fuses will withstand a 200% overload for about 20 seconds, and are used where motors or filter capacitors cause high start-up currents. They are recognized by the thick rod or coil-shaped internal structure. Fuses should be selected for a rating about 1.5 times the actual maximum operating current.

The "normal" triggering mode of an oscilloscope is *not* normal. "Auto" is the mode to use "normally." *Normal* should be called *driven*. It is used to keep the beam waiting off the screen until the trigger event occurs. And *AC Coupling* is used only when you want to filter off dc. Normally, you should use *DC Coupling,* even to view ac.

Play the odds when troubleshooting. Mechanical devices— sockets, connectors, switches, motors, relays—fail much more frequently than electronic components. Electrolytic capacitors and power semiconductors fail next most often. Other capacitors and small semiconductors fail infrequently; resistors hardly ever.

Operator error is far and away the most common cause of service calls. Have the usual operator put the equipment through its paces first, so you can spot operator errors.

8

Safety and Electrical Wiring

8.1 ELECTRIC SHOCK

This will kill you. Although it is possible for severe electric shock to actually damage body organs or render the heart nonfunctional, the most common cause of death from electric shock is the same as the cause in a drowning—the victim dies because he isn't getting air into his lungs. Electric shock most often kills by paralyzing the muscles that force air in and out of the lungs. It follows that you can save most victims of electric shock by simple mouth-to-mouth artificial respiration.

In a time of liability lawsuits and AIDS, there are legal and medical ramifications to rendering assistance to a trauma victim, as well as possible risks to your own safety. In an emergency you will have no time to decide what to do; you must weigh the consequences of acting and the consequences of not acting, and decide well in advance what you will do if you encounter a shock victim. Here are the steps you should learn if you undertake to save a victim of electric shock.

1. Learn where the main disconnect switch is. Turn off the power before touching the victim. If you cannot find the main switch, and the victim is still in contact with the power, you might attempt to pull him off with an insulated line cord or a leather belt.

2. Yell for help. Send someone to call for an emergency crew. Ask if there is anyone in the area with more emergency medical training than yourself. But do not delay—you have only two or three minutes to act.

3. A common physical reaction to shock is vomiting. If the victim is not breathing, be sure to clear the airway so this "foreign material" is not drawn into the lungs.

4. Lay the victim on his back, with head tilted back. Pinch his nose shut, place your mouth over the victim's mouth, creating an air seal. Blow gently and slowly until the victim's chest rises. To minimize risk of AIDS or other disease transfer, take care to avoid breaking the skin. Resuscitation tubes are available that make mouth-to-mouth contact unnecessary.

5. Release the nose and mouth contact, and allow the victim to exhale naturally. Repeat steps 4 and 5. Take your time. The victim doesn't need much air in his reposed state. The excitement of the emergency may tend to make you push too much air too fast.

6. Once you have begun artificial respiration, DO NOT STOP until you have been relieved by medical personnel or until you fall over from exhaustion. This is important to protect yourself legally. There are cases on record where victims have been kept alive by artificial respiration for *many* hours. Symptoms associated with death, such as stiffening of the limbs, may be observed, but only competent medical authority can pronounce the person dead.

Persons who have been trained in CPR may choose to check for heartbeat and may decide to attempt external cardiac massage. Such treatment is not likely to be successful without special training, and in the majority of cases of electric shock, it is the lungs and not the heart that need attention.

Preventing Shock. Here are some rules to help you and your associates avoid getting a shock in the first place.

1. Never work on high-voltage equipment unless there is someone else in the area to render or call for assistance in case you are knocked out by a shock.

2. Keep several shorting leads in your toolbox or work area. This is just a heavy stranded wire with insulated clips on each end. You connect it between the main voltage and ground after the power has been turned off. This will discharge high-voltage filter capacitors, which will protect you in the event that someone unwittingly turns the power back on while you are working.

3. Treat all circuits as if they are live. Don't touch metal parts unnecessarily. Keep one hand behind your back, so you will touch only one thing at a time. Remember, current needs and entry point AND an exit point to flow through your body.

4. Use insulated-handle tools. Wear special electrical insulated gloves when working on voltages above 150 V.

5. One side of the ac line, and one side of many instrument power supplies, is connected to earth ground. Avoid contact with grounded objects, since this can be the second point of contact that lets current flow through you. Don't touch water pipes, conduits, heating ducts, electrical panels, or a moist floor. Standing on a dry rubber mat will add a measure of safety. Water in a pool, sink, or tub can provide a ground contact.

6. Dry skin conducts less readily than wet skin. Keep yourself dry. Never work with electricity in a wet area. Flooded basements are death traps.

7. Remove rings, watches, bracelets, and necklaces. They can conduct current to your body; they can carry hundreds of amps themselves, turning white-hot and causing severe burns. They can get caught in machinery.

8. Guard against secondary injuries in the event that you receive a shock. Will you fall off the ladder if you make an involuntary jerk, or are you belted and positioned in such a way as to prevent a fall? If your hand jerks, will it hit something sharp or breakable, or have you planned your arm position so that a jerk will simply pull it out of the way?

9. Three-wire plugs and extension cords can prevent many shock hazards *if they are allowed to work as intended.* Never use a two-wire extension cord with a three-wire appliance. Never use an "adapter" to power a three-wire appliance from a two-wire outlet. If you notice that a three-wire appliance blows breakers when connected properly, but works when connected through a two-wire extension cord, *the appliance is deadly!* Its case is shorted to the ac hot line. Touching it and ground will give you a severe shock. Do not use it until it is properly repaired.

10. Never work with faulty electrical tools or equipment. If a plastic drill case is cracked, or the insulation on an extension cord is frayed, have it replaced. Make it a firm policy never to work with dangerous tools or equipment.

11. Ground Fault Circuit Interrupters (GFCIs) are required by law in basements, kitchens, outdoors, and on construction sites. These electronic devices sense when current is going astray (as when someone is getting a shock) and turn off the power in a fraction of a second. Never bypass them. If they are tripping it is because there is an unsafe condition. Lose your job and you may get another. Lose your life, and you get no second chance.

How much voltage is dangerous? It is easy to say that any voltage is dangerous, but no one takes special precautions when handling a 1.5-V flashlight cell, nor even a 9-V radio battery, so that is obviously not true.

1. For healthy workers, "brushing" contact with dry skin becomes dangerous above about 40 V.

2. For heavy contact from a full-hand grip, or through wet skin, open wounds, or watches or bracelets, voltages of 12 V or less must be considered dangerous.

3. For persons with medical conditions and implants such as heart pacemakers, any voltage could indeed be fatal.

4. Low-voltage sources capable of supplying currents above an amp or two can be dangerous because of possible sparks and molten metal if they are shorted out.

8.2 ELECTRICAL SAFETY

Here are further safety rules for electrical and electronics workers. Of course, safety is never encompassed by any set of rules. It is more enhanced by a constant "look-ahead" attitude: "What might happen if . . ." If something you are considering doing seems a little risky, don't say, "I'll take the chance." Figure out a safe way to do it.

1. Electric shock is *never* a joke. It is appalling to see some magazines advertising "harmless" shocker devices as a novelty. The shock may be harmless in 99 cases out of 100, but the victim in the 100th case may have a medical condition, unknown to you or even to him, that makes it fatal in his case.

2. Never over-fuse a circuit. Follow the recommended fuse size on the instrument. Occasionally, an instrument that is undergoing a slow failure will blow a proper-size fuse, but will work again if a larger size fuse is installed. *Do not use the instrument with a larger size fuse!* You will only be covering up a problem until it gets worse. "Worse" in this case may mean a catastrophic failure of the instrument, involving overheating and fire.

3. Electrical fires are termed Class-C. (Class-A involves wood, paper, fabrics, etc. Class-B involves oil, gasoline, paint, etc.) Carbon-dioxide or other gas extinguishers are best for electrical fires, because they leave little residue and are not likely to damage electronic equipment. The popular dry chemical extinguishers may be used on Class-C fires, but may leave damaging residue in the instruments hit by the stream. *Never use water* on an electrical fire. Use of foam-type extinguishers on live electrical circuits may extend the shock hazard.

4. Automobile battery explosions are much more common than most people realize. Charging a wet-cell battery creates highly explosive hydrogen gas. Never charge a car battery in an enclosed area. Give it plenty of circulation to open air. Never permit a spark or flame near a charging battery. When jump-starting a car,

connect the red (hot) wires first. Make the last (ground) connection to the good battery (not the charging one) at a point on the vehicle chassis far away from the battery. Do not make this last connection at the battery post. Wear eye protection and turn your face away to protect yourself in the event of an explosion.

5. If a ring, watch, or tool connects the main +12 V wire to a car chassis, hundreds of amps will flow and the metal object may melt, throwing hot metal on your skin and eyes. Take rings and watches off, and guard against tools shorting the +12 V to ground.

6. Never solder when the work is above your head. Get the work lower, or your head higher so hot solder won't drip on you. Point the wire end down at the floor or workbench when cutting or stripping, so the pieces won't fly up into somebody's face. Wear eye protection when soldering or when cutting or stripping wire.

7. Provide adequate ventilation in areas where people are soldering or working with printed circuit chemicals or painting.

8. There is an art to climbing towers and working on roofs. Don't attempt to do this unless you have the proper training and safety equipment. Climbing a tower without a safety belt is foolish and extremely dangerous.

9. Never work around machinery, or allow others to come near machinery, while wearing neckties, sweaters, loose clothing, or long hair. Drill presses, mills, and lathes can do hideous things to people who get sweaters or their hair caught in them.

8.3 ELECTRICAL WIRING PRACTICE

This section is meant to familiarize the technician with some of the more basic and commonly abused electrical-wiring practices. It is not intended to substitute for the supervision of a qualified and licensed electrician. For complete data on this subject, consult the *National Electrical Code* book, published by the National Fire Protection Association, 470 Atlantic Avenue, Boston, MA 02210, and your local building inspector.

A wiring plan for a residence or small commercial building is shown in Figure. 8.1 on page 264. In larger systems there may be several branch breaker boxes located remotely from the service entrance and connected to it by feeder lines.

Equipment grounding. All new or replacement wiring should include an equipment-grounding wire (commonly called a ground wire) of a capacity to conduct safely any fault current likely to be imposed. This wire may be bare, or it may have green or green-and-yellow striped insulation. Terminals to be connected to this ground are colored green or are marked with the word *green*. Equipment grounding can be achieved by attaching (bonding) from the service-entrance neutral wire to the street side of the city water meter or to a ground rod not less than 8 feet in length.

The function of the equipment-grounding wire is illustrated in Figure 8.2 on page 265. Without this third wire, a short from the appliance or instrument case to the live wire places the case at 115 V to earth ground and presents a shock hazard. The grounding wire keeps the case at earth potential and shorts the entire circuit in the event of a hot-wire-to-case short, thus tripping the circuit breaker.

Bonding must be made between the grounding wire and all conduits, metal boxes, and metal equipment cases.

When replacing outlets or wiring in an existing two-wire system, the grounding terminal of the new outlet must be connected to a cold-water pipe or other earth-grounded pipe. It must *not* be connected to the existing neutral wire, even though this wire is at ground potential. Figure 8.3, on page 265, shows how this error places the equipment case at 115 V (through the relatively low motor resistance) in the event of an open in the neutral conductor.

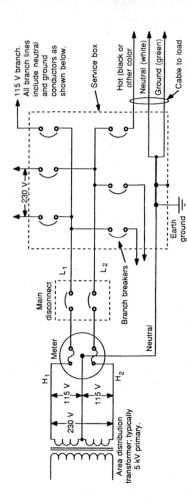

Figure 8.1 Typical wiring for electrical distribution in a home or small commercial building.

264

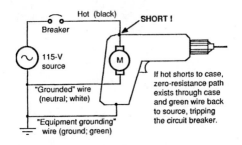

Figure 8.2 The third wire to an electric appliance causes the breaker to open if the device is unsafe.

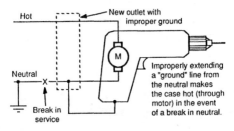

Figure 8.3 The third wire must have an independent ground, even though the neutral wire is grounded.

Current-carrying conductors. The grounded conductor (commonly called the neutral or cold side) should have white or gray insulation. Terminals to be connected to neutral should be plated with a white or silver metal, or should be marked with the word *white*.

The ungrounded conductor(s) (commonly called the live or hot side) may have any color insulation other than green, gray, or white. Black is commonly used, with red as the first alternative. Terminals to be connected to the live wire are generally yellow or copper colored.

Electrical wiring circuits. Switches, fuses, and breakers should always be placed in the live side, never in the neutral side of a circuit. Reversing this rule would leave hazardous voltage at the load (to earth ground) even with the switch off, and could place unbalanced loads across the 230-V line, resulting in damage to the lower-power load, as illustrated in Figure 8.4, below.

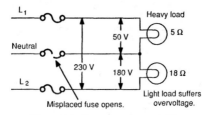

Figure 8.4 Never fuse the neutral line.

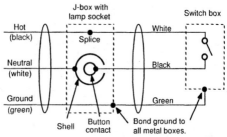

Figure 8.5 Physical layout of wiring for a simple electric light switch.

The shell of a lamp socket is connected to the neutral side (button to hot side) to minimize shock hazard by accidental contact with the shell. Figure 8.5, above, shows the wiring layout for a simple ceiling light and switch. Figure 8.6, on the next page, shows the layout for two-point control of a single light (the so-called three-way system).

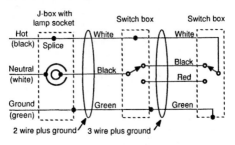

Figure 8.6 Two-point control of a single light.

A **ground-fault interrupter** (GFI) is an inexpensive electronic device that compares the live-wire and neutral-wire currents, and trips a breaker if they are unequal. Since most electric shock involves a current path from the live wire to an independent ground (not the neutral wire), the GFI can save a potential shock victim by turning off the power in a matter of milliseconds. Ground-fault protection is required in 115-V outlets installed outside, in a bathroom, in a garage, and on temporary outlets at construction sites. Local codes may also require GFIs in basement and kitchen branch circuits.

Load calculation. Dwelling rooms should have outlets placed so that no point on any wall is more than 6 ft. horizontally from an outlet. Wall sections 2 or more ft wide should be included in this determination. At least one outlet should be provided

- At any kitchen countertop more than 1-ft wide
- Adjacent to the bathroom basin
- In the basement
- In an attached garage
- Outdoors

The number of branch circuits required for general lighting and convenience outlets in dwellings, stores, and shops may be determined by allowing 3 watts per square foot of habitable floor space and apportioning the load equally among 15-A branches. For example,

267

$$1400 \text{ ft}^2 \times 3 \text{ W/ft}^2 = 4200 \text{ W}$$
$$4200 \text{ W} \div 115 \text{ V} = 36.5 \text{ A}$$

A *minimum* of three 15-A branch circuits is therefore required for a lighting load. Local codes may require more.

A minimum of two 20-A branch circuits are required exclusively for outlets provided for small appliances in the kitchen-pantry-dining area. Also, at least one 20-A branch circuit is required for receptacles in the laundry area.

Separate branch circuits should also be provided for all fixed appliances, such as ranges, ovens, air conditioners, dryers, water heaters, space heaters, and heat pumps. Branch-circuit ratings should equal or exceed the nameplate ampere or kVA rating of the appliances.

Wiring practice. Conductors should be sized so that the maximum voltage drop to the farthest outlet cannot exceed 3% from the branch circuit breaker, nor 5% from the feeder circuit breakers. Good practice in dwellings is to use No. 14 copper wire for 15-A branch circuits and No. 12 copper wire for 20-A branch circuits.

Where additional outlets and lighting are to be installed, as in building additions or in finishing rooms in basements or attics, it is necessary to install one or more new circuit breakers at the branch-service point and wire the new branches from this point. It is hazardous and generally illegal simply to extend new outlets and lighting from the nearest existing branch, unless load calculations have shown that the maximum branch current will not be exceeded thereby.

Outlets should not be placed over the tub in a bathroom, as this is an invitation to place an appliance where it can fall into the bather's lap.

Wires should be spliced or joined only inside metal boxes, and then they should be joined with wire nuts or solder. Twist-and-tape connections are certain to corrode with time, resulting in high resistance, overheating, and possibly fire.

When terminating three-wire flexible cords, it is a good idea to leave the grounding wire (green) a little longer than the live and neutral wires so that it will be the last to sever if the cord is strained.

9

Dictionary of Computer and Electronics Terms

ab, *obsolete*. Prefix for electrical units formed in the cgs (centimeter-gram-second) system.

Absolute, *mathematics*. The value of a number without regard to sign, as opposed to *algebraic* value. The absolute value of –7 is 7. *Physics*. The value of a quantity measured on a scale with the zero reference at zero value. For example, *gage* pressure is pressure above atmospheric (14.7 lb/in^2); *absolute* pressure is pressure above zero lb/in^2. *Computers*. An addressing mode in which the operand address is specified without other references or modifiers.

Acceptor impurity, *semiconductors*. An impurity element containing three electrons in the outer shell, leaving the fourth position vacant to accept a free electron.

Accumulator, **1.** *computers*. A register that stores the result of a data-move, arithmetic, or logic operation.
2, *obsolete*. Storage battery.

ACL, *internet*. Access Control List. A scheme for defining who is granted access to particular parts of a web site.

Active component. A component capable of voltage or current gain, or switching.

Active transducer. A transducer producing voltage or current without an external source of electric energy.

ActiveX, *internet*. A technology that permits animated segments on web pages.

Ad hoc. Literally, *to this*. A style of writing computer programs just as it comes naturally, without structure or restrictions. Less favorably called *spaghetti code* or *rat's-nest programming*. See *linear structured*.

A-to-D converter. Analog-to-digital converter (ADC).

Ada. Standard programming language of the U.S. defense department. Similar to Pascal. Named for Ada, Countess Lovelace; assistant to Charles Babbage, inventor of the digital computer (1833.)

Address, *computers.* Numerical designation unique to each particular word in memory.

Addressing modes. The various methods by which a computer obtains or dispatches the data manipulated by a given instruction. See: Absolute, Direct, Immediate, Implied, Indexed, Indirect, Relative, and Zero-page.

Admittance. The reciprocal of impedance. Quantity symbol: Y. Unit: Siemens (S).

AFC. Automatic Frequency Control, used in FM radio receivers to avoid the necessity for continuous retuning.

AGC. Automatic gain control; common in radio and TV receivers.

AIX. IBM's implementation of Unix.

ALC. Automatic level control; used in SSB transmitters to keep output level constant.

ALGOL. ALGOrithmic Language; a compiler language.

Algorithm. The specific procedure selected for solving a given problem.

Alias, 1. Spurious signal obtained in digital recording of analog data when signal frequencies (f_o) above one-half the sampling rate (f_s) are allowed into the A-to-D converter. Alias frequency is $f_s - f_o$. **2.** In Apple's System 7, a file whose sole purpose is to represent another file.

Alpha (α). The ratio of collector current to emitter current in a bipolar transistor. Also referred to as h_{fb}.

Alpha cutoff (f_α or f_{hfb}). The frequency at which alpha drops to 0.707 of its low-frequency value.

Alpha particle. A particle consisting of two protons and two neutrons.

ALS. Advanced Low-power Schottky. A fast family of TTL integrated circuits.

Alternator. An ac generator.

ALU, *computers.* Arithmetic Logic Unit.

AM. Amplitude Modulation. A radio transmission technique in which the strength (amplitude) of the transmitted signal is varied in accordance with the intelligence (generally voice or music) being transmitted.

Ambient, *adjective*. Associated with a given environment.

Analog. Representation of one quantity by means of a second quantity which is proportional to the first. For example, on a clock face, *time* is understood to be proportional to *angle* of the hands of the clock.

Anode. Positive electrode.

Anonymous FTP, *internet*. A file transfer in which data is downloaded or placed without requiring the operator to identify himself.

ANSI. American National Standards Institute.

API. Application Programming Interface.

Apparent power. The product of voltage times current in an ac circuit, regardless of whether V and I are in phase. In reactive circuits, V and I are not in phase, and some of the energy supplied to the circuit on one quarter cycle is returned to the source on the next quarter cycle.

Applets, *internet*. Short application programs downloaded from a network as they are needed, and runnable on any computer accessing the network.

Application, *computers*. A program designed to perform a function for the end user, such as a word processor, a drawing program, or a spreadsheet.

Archie, *internet*. An internet search program.

Architecture, *computers*. The conceptual design of computer hardware, including number of accumulators, word length, addressing modes, and similar features.

Archive. A storage of infrequently used or historical data.

Arithmetic shift, *computers*. Shift of digit positions in a word, resulting in multiplication or division of the number by a power of the number-system base.

Armature. The moving part of a magnetic device, such as a motor or relay.

Armature reaction. Distortion of the magnetic field in a motor or generator caused by armature current.

Armstrong oscillator. An oscillator using a transformer to feed part of the output of an amplifier back to the input.

ARP, *internet*. Address Resolution Protocol. The protocol for mapping IP addresses to physical addresses.

Array. A listing of data by rows and columns.

ASCII (*say* ASK–ee.) American Standard Code for Information Interchange. A six- or seven-bit digital code representing English letters, numbers, and punctuation marks. *The letters* II *are sometimes mistakenly read as a Roman numeral; however, there is no "2" involved.*

ASIC. Application-Specific Integrated Circuit. An IC designed for a unique application by the end user on a computer simulator, as opposed to one designed by the manufacturer for sale to a more general market.

Aspect ratio. The ratio of width to height for a visual display. Equal to 4/3 for NTSC television video.

Assembler. A computer program that converts user-recognizable coding into machine code on a one-instruction to one-instruction basis, also keeping track of data locations by data names.

ATE. Automated Test Equipment.

ATM, *internet.* Asynchronous Transfer Mode.

Audio-taper potentiometer. A potentiometer in which resistance changes slowly at counterclockwise rotation and rapidly at clockwise rotation to make response seem linear to the human ear.

AVC. Automatic Volume Control. Commonly employed in radio receivers. Also called AGC.

AWG. American Wire Gage. See Chapter 2.

Axial leads. Component leads coming out each end of a tubular component, allowing horizontal mounting to a circuit board. See *Radial leads.*

Ayrton Shunt. Ammeter shunt resistors switched to provide multiple ranges while keeping the same resistance across the meter for optimum damping and not exposing the meter to full current during switch transitions.

Back diode. A tunnel diode with very low peak tunnel current, used as a low-voltage rectifier conductive in the reverse direction.

Background, *computers.* A computer operation that takes place without being visible on the monitor, and without continuous control from the keyboard.

Backplane, *computers.* Wire interconnections between circuit-board edge connectors.

Ballast resistor. A resistor that increases in value as current increases, tending to keep current constant.

Balun. BALanced-to-UNbalanced transformer between twin-lead and coaxial lines.

Bandwidth. The span of frequencies between the lower −3 dB cutoff and the upper −3 dB cutoff of a system.

Bar Code. The rectangular pattern of bars representing the 10-digit Universal Product Code (UPC) for a retail-sales item. Each digit consists of two dark bars and two white spaces one through four elements wide.

Barrier strip. An insulative strip containing a row of screw terminals for connecting an instrument to a cable.

BASIC, *computers*. Beginner's All-purpose Symbolic Instruction Code. A versatile and easy-to-use high-level interpreter language, also available as a compiler.

Baud. Serial data transmission rate in bits per second. A rate of 1000 baud represents the possibility of 1000 level changes per second, or a square wave of 500 Hz.

BBS, *internet*. Bulletin Board System. A service by which users with common interests can post messages and replies on the internet.

BCD. Binary Coded Decimal.

Benchmark, *computers*. A short program designed to furnish a basis for comparison of two computers or similar application programs, usually on the criterion of speed.

Beta (β). The ratio of transistor collector current to base current. Also termed h_{fe} .

Beta version. Early production versions of a product shipped by manufacturers to trusted customers to gauge customer reaction and weed out problems.

BFO. Beat-Frequency Oscillator; used in receivers to heterodyne with code or SSB signals to produce audible frequencies.

BGA. Ball Grid Array. A package-pin arrangement for surface-mount electronic components in which contacts from the chip are brought out to a grid of contact balls on the bottom of the package. The balls are then soldered to the surface of the circuit board.

Bias. Voltages or currents required by an electronic component for its proper operation. Transistors generally require dc bias; some transducers, such as recording tape, may use ac bias.

Bidirectional data bus. Interconnections between two or more digital devices in which data can be written into or read from each device via the same set of lines.

Bifilar. Two wires or filaments wound side-by-side.

Bifurcated contact. A contact split into two fingers for redundant contact.

Binary. Having only two possible states; for example: *true* and *false*, 1 and Ø, or HALT and $\overline{HALT}$.

BIOS, *computers*. Basic Input/Output System. A part of the operating-system program.

Bipolar transistor. The common NPN or PNP transistor, which uses both positive and negative charge carriers in its main current path. Also called *bijunction*.

Bit. Binary digIT. The contents of a single flip-flop or memory cell. Represented by a 1 or a Ø.

Bit-sliced. A technique of building long-word-length computers by paralleling shorter-word-length processors.

BJT. BiJunction Transistor. A bipolar transistor.

BNC. A popular coaxial connector, operated by a *push and twist* motion. Letters *B* and *N* refer to the designers.

Boffin, *British*. A theoretical engineer or researcher.

Bomb, *computers*. Total failure of a computer program or operating system, necessitating a *cold boot*.

Boot, *computers*. To start up a computer.

Bootstrap, 1. *computers*. A short series of instruction codes that program the machine to read the following codes from its input devices. (A machine with *no* program in memory isn't even capable of reading its input devices.) Named from the adage "to pick yourself up by your own bootstraps." **2.** *electronics*. A technique for increasing input impedance by feeding a portion of the output voltage back to buck the input voltage.

bps. Bits Per Second. A data transmission rate.

Branch instruction, *computers*. An instruction that alters the normal address-counting sequence of instructions if specified conditions are met.

Breakout box. A test device that brings all the wires of a cable or connector out to labeled and easily accessible terminals for ease in checking the signals.

Breakpoint, *computers*. An address that, when reached, will cause the computer to halt so that its register and memory contents can be examined in troubleshooting.

Browse, *internet.* To search portions of the net in a relatively non-directed way, looking for items of interest. Also *surfing* or *cruising.*

BTW, *internet.* By The Way.

Bubble memory. A nonvolatile serial-access integrated-circuit memory using magnetic domains for storage.

Bubble sort, *computers.* A procedure for sorting a group of numbers by continuously exchanging pairs of numbers, placing the higher number later in the list. Commonly used as a teaching exercise.

Buffer, 1. A circuit that increases the driving capability of a signal. 2. *computers.* A memory location for temporary storage of data. 3. *computer printers.* A device, consisting mostly of memory, that feeds data to the printer, thus freeing the computer for other tasks.

Bug. A flaw in a program or hardware system.

Bundled. A marketing term denoting different items sold as part of a package.

Bus, 1. A line or conductor that distributes a voltage or signal (or ground connection) to a number of points.
2. *computer.* A group of lines acting together; i.e., the *data bus, address bus,* and sometimes *control bus.*

BX cable. Insulated wires in flexible metal armor.

Byte. Standard binary data unit consisting of eight bits.

Cache, *computers.* A small, fast-access memory for storing frequently used data and instructions.

CAD, CADD. Computer Assisted Design/Drafting.

CAE. Computer Assisted Engineering.

CAI. Computer Assisted Instruction.

Calculator. A device that performs mathematical operations in response to instructions and data entered manually, automatically, or from a storage device.

CAM. Computer Assisted Manufacturing.

Capacitance. The property of being able to store charge. Measured in Farads (F), and calculated as $C = Q/V$.

Capacity. A rating of a battery's current-delivering capability, expressed in ampere-hours (A·hr).

Carrier system. A system for sending a number of voice or data channels over a single wire by modulating a number of higher-frequency carrier signals.

Cascade. A number of amplifier stages connected output-to-input

Cascode. A common-emitter stage feeding a common-base stage, or a common-cathode stage feeding a grounded-grid stage.

Cathode. Negative electrode.

CATV. Community Antenna Television.

CCD. Charge-Coupled Device.

CCITT, *internet*. Consulting Committee for International Telegraph and Telephone. A committee under the ITU for data communications standardization.

CD, 1. Compact disk. A rotatable disk containing a large amount of read-only data, and read by a laser beam.
2. Capacitive discharge.

CD ROM. A read-only memory device with large storage capacity, using a compact disk.

CDPD, *internet*. Cellular Digital Packet Data. A standard for sending internet data via cellular phones.

Cellular. A mobile-radio communications system in which the mobile units are switched, under computer control, to various fixed relay stations (cells) as the mobile unit roves out of range of the original cell.

CGA. Color Graphics Adapter. A video output system for IBM PC and compatible computers. Best text mode is 640×200 pixels. Best graphics mode is 640×200 in 2-color mode, 320×200 in 4-color mode.

CGI, *internet*. Common Gateway Interface.

Checksum. A number obtained by adding all previous numbers in a list of numbers and discarding any higher-order digits thus generated, or any similar technique for deriving a number from a list of numbers. Used to check the accuracy of the list.

Choke. An inductor used to block the flow of ac above a specified frequency.

Chopper. A device for converting continuous light or voltage to ac by regular interruption. Used to permit the signals to be handled by ac-coupled amplifiers.

CIM. Computer Integrated Manufacturing.

CISC, *computers*. Complex Instruction Set Computer, typified by the 486 and pentium chips. See *RISC*.

Clamper. A circuit used to re-establish a dc level on an ac waveform.

Class-A amplifier. An amplifier in which collector or plate current flows for the entire 360° of the signal cycle. Produces essentially undistorted outputs with poor power efficiency.

Class-B amplifier. An amplifier in which collector or plate current flows for one half of the 360° signal cycle. Produces half-wave outputs with better power efficiency.

Class-C amplifier. An amplifier in which the transistor or tube is switched from saturation to cutoff as rapidly as possible to avoid power waste in the device. Produces distorted pulse outputs with high power efficiency.

Clock, *computers.* A pulse generator with which data transfers are generally kept in step.

CML. Current-Mode Logic. Also called ECL.

CMOS. Complementary Metal-Oxide Semiconductor. A low-power logic family using field-effect transistors.

Coaxial. Cables or connectors having a center conductor completely surrounded by an outer conductor, placing the two conductors on the same *axis.* Coax.

COBOL. COmmon Business-Oriented Language. A popular compiler language.

Coherent radiation. Radiation of a single frequency having definite phase relationships between parts of the wave.

Cold boot, *computers.* Restarting a computer by turning the power off and back on again. Opposed to *warm boot.*

Colpitts oscillator. An oscillator using a phase-shifting feedback network consisting of an inductor and two capacitors.

Comb filter. A filter that rejects signals at certain regular intervals of frequency, so that its frequency-response curve looks like a comb.

Common mode. Applied to both sides of a balanced line or amplifier simultaneously.

Common Mode Rejection Ratio (CMRR). Ratio of amplifier gain for differentially applied signals to its gain for signals applied in common to the two inputs.

Compand. A technique used in digital recording of analog data to achieve a higher dynamic range. An increase in count at high levels represents a larger voltage change than the same increase in count at low levels.

Compiler. A computer program that translates a complete source-code list, consisting of user-recognizable instructions, into machine code, with one user instruction generally producing many machine instructions.

Composite video. A signal (bandwidth about 5 MHz in US broadcast TV) that contains black-and-white, color (if applicable), and synchronizing information.

Computer. A machine that performs mathematical and logic operations according to instructions stored in memory, and alters its instructions, or branches to new instructions, depending on intermediate results.

Concatenate. To place two or more short data words together, thus producing a single longer data word.

Condenser, *obsolete*. Capacitor.

Conductance. The reciprocal of resistance; the ratio of current to voltage. Quantity symbol: G. Expressed in the unit *Siemens*.

Contact potential. A millivoltage appearing across the junction of two dissimilar metals and varying with temperature.

Control unit. That section of a computer which interprets the binary code in the instruction register to activate selected operational functions of other registers or the ALU, such as adding, incrementing, writing, or ANDing.

Core storage, *computers*. Fast read/writable memory consisting of one tiny saturable magnetic element per bit. Since the technique is now largely obsolete, the term is sometimes loosely used to refer to the computer's RAM.

CP/M. Control Program for Microprocessors. An early operating system for small systems using the 8080 microprocessor, characterized by commands consisting of one to four keystrokes. Forerunner of *DOS*.

cps, **1**. Characters Per Second; a measure of printer speed. **2**. *obsolete*. Cycles Per Second; replaced by Hz, *Hertz*.

CPU. The Central Processing Unit of a computer.

CRC, *computers*. Cyclical Redundancy Check. A check-sum technique for catching and correcting errors in a binary data record.

CRO. Cathode-Ray Oscilloscope.

Cross assembler, *computers*. An assembler that runs on one machine (the *host* processor) but produces code that will run on another machine (the *target* processor.)

CRT. Cathode-Ray Tube; the common TV and video display device.

Cryogenics. The science of material behavior at extremely cold temperatures.

Cryptography. The science of encoding information to prevent unauthorized access.

CSP. Chip Scale Package. A technique of reducing an integrated-circuit's package to a size only slightly larger than the chip itself.

CTCSS. Continuous Tone Coded Squelch System. A system by which a radio transmitter sends a subaudible tone (67 to 193 Hz) to activate only receivers set to receive the same tone. Also called *PL* (Private Line.)

Curie point. The temperature above which a magnetic material loses its magnetic properties.

Current mirror. A technique used in integrated circuits in which current in an active transistor is held equal to current in a second transistor by virtue of identical base-emitter voltages.

Current Mode Logic. A family of solid-state logic in which a constant current is switched to one of two transistors to represent the two logic states. Its non-saturating character makes it extremely fast. Also called *emitter coupled logic.*

Cursor, *computers.* A character, such as a flashing line or box, that indicates the position on a display screen where key entries will appear.

Cutoff. The zero-current condition of a transistor or switching device.

Cybernetics. The study of computer-based automation and artificial intelligence, especially as it mimics or interacts with human intelligence.

Cyberspace, *internet.* A general term referencing the activity and culture current among users of the internet.

D-to-A converter (DAC). Digital-to-Analog Converter.

DAC. Data Acquisition and Control. Use of computers to sense and record data, and control an industrial process.

Daisy chain. A wiring structure in which the wire is strung from the feed point to the first terminal, and continues to the next, and to subsequent terminals; as opposed to the *tree* structure, in which a separate wire runs from the feed point to each terminal.

Daisy wheel. A type of impact printer in which the fully formed characters appear at the ends of "spokes" on a wheel. Characterized by excellent quality print but slow speed.

Damp. To make an oscillation come to rest.

Darlington compound. Two bipolar transistors with the emitter of the first feeding the base of the second, and the collectors tied together.

DARPA. Defense Advanced Research Projects Agency.

DAS, 1. Digital Analysis System. **2.** Data Acquisition System.

DAT. Digital Audio Tape.

Data base, *computers*. An application program designed to function somewhat like a card file, providing storage, access, and sorting of data by different criteria.

DB-25. A popular 25-pin connector used with desktop computers.

DCC, *internet*. Direct Client-to-Client. A technique allowing IRC users to exchange data packets.

DCE. Distributed Computing Environment.

Decoupling. Filtering, usually with capacitors and inductors or resistors, to prevent signals from passing from one circuit to another.

Dedicated computer. A computer programmed for and often wired for a specific task and not intended for other tasks; in contrast to a *general-purpose* computer.

Degauss. To remove residual permanent magnetization.

Delimiter. An alphanumeric character, often a slash, placed at the beginning and end of a field of text thus marked for special treatment by a computer program.

Depletion zone. A portion of a semiconductor near the junction void of charge carriers.

Descender. The portion of a character below the base line in such letters as g, j, p, q, and y.

Device Driver. Software that acts as an intermediary between a CPU and a peripheral device.

Diac. A two-terminal semiconductor device that conducts abruptly at a specified breakdown voltage in either direction.

Dial-up IP account, *internet*. Internet access with a modem and phone line via an intermediate computer, using TCP/IP protocol and allowing full access to the net. Comes in PPP and SLIP modes. See, for comparison, *Shell* account.

Differential. Applied between the two sides of a balanced line or amplifier.

Diffused junction. A P-N junction formed by diffusion of P-type impurity from a vapor into a region of N-type semiconductor, or vice versa.

Digital. A system in which codes or characters are used to represent numbers or physical quantities in discrete steps; as opposed to *analog*.

Digitizer. A circuit or component that causes an analog quantity to be represented in digital form.

DIN. German Industrial Standard.

DIP. Dual In-line Package. Commonly used for integrated circuits.

Diplexer. A coupler that allows two transmitters or receivers to operate on a single antenna.

Dipole. An antenna one-half wavelength long and split at its center for connection to a feed line.

Direct addressing, *computers*. An instruction mode in which the address of the data to be operated upon is specified by the word or words following that instruction.

Direct coupling. A signal path between two stages, consisting of a conductor, resistor, or battery.

Disk, *computers*. A rotatable disk coated with magnetic material and used for bulk storage of data and programs.

Dithering. A technique of achieving an intermediate position by rapidly switching between two positions on either side.

DMA, *computers*. Direct Memory Access. A technique for high-speed memory reading or writing, bypassing the processor.

DMM. Digital MultiMeter. A meter with voltage, current, and resistance-measuring functions.

DNS, *internet*. Domain Name Service.

Documentation. Instructions, notes, and diagrams prepared to assist in understanding and use of an instrument or computer program.

Domain name, *internet*. The name of the specific computer being accessed. Last letters indicate: **edu**cational, **com**mercial, **gov**ernment, non-profit **org**anization , or public **net**work.

Donor impurity. An impurity that increases the number of free electrons in a semiconductor.

Doping. Addition of impurities to a semiconductor.

DOS. Disk Operating System. A program that enables a computer to read and write to its disk drive, and otherwise manage data files.

Dot matrix. A type of printer in which characters are formed by individual dots, produced by tiny pins impacting the paper through an inked ribbon.

Double-base diode, *obsolete term*. A unijunction transistor.

Double-density disk. A technique for nearly doubling the amount of information stored on computer floppy disks, generally by eliminating unnecessary clock pulses.

Double-precision arithmetic, *computers*. Use of two words rather than one to represent each number.

Download, *computers*. To transfer data or programs from a larger system to a smaller system.

DRAM. See *Dynamic RAM*.

Driver. The amplifier stage preceding the output stage.

DSO. Digital Storage Oscilloscope.

DSP. Digital Signal Processor.

DTE. Data Terminal Equipment.

DTMF. Dual Tone, Multi-Frequency. The standard telephone "touch tone" frequency pairs.

DTP. DeskTop Publishing.

DTR. Data Transfer Rate.

Dumb terminal. A computer terminal consisting of a keyboard and monitor, but having no local computer processing capability.

Dump, *computers*. To transfer the contents of memory into another storage device.

Duplexer. A rapid-switching device that permits use of the same antenna for transmitting and receiving.

Duplex operation. Transmitting and receiving without noticeable switching between send and receive periods. In serial data communication, *full duplex* refers to simultaneous two-way transmission, and *half duplex* refers to rapid switching between directions.

DUT. Device Under Test.

Duty factor. Ratio of working time to total time.

DVM. Digital VoltMeter.

DVORAK. A non-standard typewriter keyboard arranged so the most-used letters are activated by the strongest fingers. See *QWERTY*.

DX, *radio.* Long distance.

DX2. A computer CPU with an external bus speed half that of the internal bus.

Dynamic. Changing, or in motion; ac.

Dynamic braking. Slowing an electric motor by connecting it as a generator feeding energy to a resistance or voltage source.

Dynamic loudspeaker. A speaker in which the audio current is carried by a coil that moves with the diaphragm.

Dynamic RAM (DRAM). Semiconductor memory in which data will be lost if not rewritten periodically, typically every 3 ms.

Dynamic range. In an amplifier, the amplitude range between the lowest signal level that will not be deteriorated by noise and the highest level that will not be deteriorated by overdriving the system.

Dynamic resistance. The ratio of a *change* in voltage to a *change* in current; as opposed to *static* or *bulk* resistance.

Dynamo. An electric generator.

Dynamotor. A rotating machine with two windings on one armature for converting one type of power to another, such as 12 V dc to 400 V dc.

e. The base of natural logarithms. 2.71828....

E, EE, EXP. Designations used on some pocket calculators and computer programs to indicate powers of ten. For example, 2 E 4 indicates 2×10^4.

EAROM. Electrically Alterable Read Only Memory. A semiconductor memory in which data can be written word-by-word, but more slowly than it can be read. Also called *read-mostly memory.*

Earth (*British*). Electrical ground. Also used as a verb.

ECAD. Electronic Computer Assisted Drafting.

Echo, 1. In electronic signaling, the reflection of a signal caused by a sudden change in the impedance of the transmission medium. **2.** An error detection scheme in which data is sent to a distant node that is obligated to send the data back immediately.

ECL. Emitter Coupled Logic. A high-speed logic family, also called CML.

ECM. Electronic CounterMeasures.

ECO, 1. Electron-Coupled Oscillator. A vacuum-tube oscillator using a tuned circuit in the screen grid. **2**. Engineering Change Order.

Editor. A program used to create and modify data files, especially computer source-code files.

EDP. Electronic Data Processing.

EEPROM. Electrically Erasable Programmable Read Only Memory. A semiconductor memory that can be bulk-erased and rewritten without removing it from the target system.

Effective Radiated Power (ERP). The product of antenna input power and antenna gain.

EGA. Enhanced Graphics Adapter. A video output system for IBM PC and compatible computers. Best text mode is 640×350 pixels. Best graphics mode is 640×350 with 64 colors.

EHF. Extremely High Frequencies. The radio spectrum from 30 GHz to 300 GHz.

EISA, *computers* (say *EE-sah*). Extended Industry-Standard Architecture. A bus structure for advanced 80386-based computers, retaining compatibility with earlier 80286-based systems.

Electric, *adjective*. Using or activated by current in a conductor; as *electric motor*.

Electrical, *adjective*. Related to electric devices but not operated by electric current; as *electrical engineer*.

Electricity, 1. *noun*. A property of electrons and protons, expressed numerically as charge in coulombs. **2**. *adjective*. Electrical; as *electricity book*.

Electronic, *adjective*. Using or activated by electric current in semiconductors or evacuated chambers; as *electronic computer*.

Electronics, 1. *noun*. The field of technology that deals with electronic devices. **2**. *adjective*. Related to the field of electronics; as *electronics technician*.

Elegant. Said of a computer program or electronic design that solves a problem in a remarkably simple or straightforward way.

Embedded software. Code not visible to the end user.

EMF. ElectroMotive Force. Electric potential; voltage.

EMI. ElectroMagnetic Interference.

Emoticons, *internet.* Icons built of ASCII characters to show emotions when communicating via internet:

:-)	smile	:-(	sad
8-)	wearing glasses	;-)	a wink
:-&	tongue tied	:-o	shocked
:-p	tongue sticking out		

Emulator, *computers.* A general-purpose computer used in developing dedicated computer systems; capable of running the target-system program and simulating various inputs and outputs.

Enhancement-mode FET. An insulated-gate FET that is nonconductive at zero gate-source voltage and turns *on* with forward gate bias.

Envelope. The overall shape of an amplitude-modulated waveform, disregarding individual rf variations.

Epitaxial layer. A thin semiconductor layer condensed from a vapor (grown) onto a thicker substrate layer; used to obtain high purity and regular crystal structure.

EPROM. Erasable Programmable Read Only Memory. A semiconductor memory into which data can be written only by a special programming fixture, and which can be bulk-erased by exposure to ultraviolet light.

ESD. ElectroStatic Discharge or ElectroStatic Damage. Damage to semiconductor devices, especially MOS types, from static charges accumulated on nonconductive articles in the environment.

ESR. Equivalent Series Resistance. A resistance conceived of as being in series with a capacitor, which accounts for the power lost when ac flows through it.

Ethernet. A LAN system operating up to 10 Mbps. In common usage, "Ethernet" refers either to the DIX (DEC - Intel -Xerox) version of this specification or to the IEEE version, more formally known as "802.3."

Exclusive OR. A two-input logic function whose output is true if one input OR the other is true, but not if both are true.

Executive program. A program that manages the loading, running, and outputting of other programs.

Extrinsic semiconductor. A semiconductor whose electrical properties are determined by the added impurities.

Fail Soft. A computer program that avoids catastrophe in the event of a system failure by taking some appropriate action. The "limp home" mode in automotive engine controllers is an example.

Fan out. In digital logic, the number of logic-gate inputs that can be driven by one output of the same family.

FAQ, *internet*. Frequently Asked Question.

Faraday shield. An electrostatic shield that passes magnetic and/or electromagnetic fields.

FAT. File Allocation Table. A portion of a computer disk listing where various files may be accessed.

FAX. Facsimile. Remote transmission of images from a paper original to a paper copy.

FEA, FEM. Finite Element Analysis, Modeling. A computer technique for analyzing systems by breaking them into a very large (but finite) number of very small elements and summing the effects on all elements together. A simulation of the classical calculus technique, which uses an infinite number of elements.

Female connector. One for which the main conductor or center wire is a receptacle.

Ferrite. A powdered magnetic material compressed and bonded into a desired shape; used as a core for inductors; capable of high Q at high frequencies.

FET. Field Effect Transistor. A semiconductor amplifying and switching device in which current through a single-polarity *channel* is controlled by the voltage applied to a reverse-biased *gate* electrode. Characterized by a very high input resistance.

Fetch, *computers*. To retrieve a piece of data from memory.

FFT. Fast Fourier Transform. A computerized circuit-analysis algorithm.

Field, *television*. The entire area of the display, although not necessarily completely filled. See *Frame*.

FIFO. First In, First Out. A system of storing and retrieving data in a stack memory.

File, *computers*. A block of related data.

Firewall, *internet*. A security system placed on a local area network to protect it from outside access.

Firmware, *computers*. Programs supplied with a computer as part of its operating system and usually stored in nonvolatile memory.

Flag, *computers.* A single bit of data set or cleared to indicate the presence or absence of a particular condition.

Flame, *internet.* An angry or abusive message sent via internet.

Flip flop. A digital circuit generally consisting of two cross-connected transistors, in which one transistor will remain turned on while the other remains turned off, until an input signal reverses these two states. Bistable multivibrator.

Floating-point arithmetic. Arithmetic operations in which the position of the decimal point is variable.

Floating state. A semiconductor logic output that pulls neither to a high nor a low voltage state.

Floppy disk, *computers.* An easily portable magnetic storage device.

FLOPS, *computer.* FLoating-point OPerations per Second. A measure for comparing computers on the basis of speed.

Flow solder. Mass soldering of printed-circuit boards by moving them over a wave of molten solder.

Fluorescence. Emission of light by a substance when exposed to radiation or impact of particles; ceases within a few nanoseconds of bombardment.

Flutter. Distortion of sound caused by speed variation of tape or disk during recording or playback.

FM. Frequency Modulation. A radio transmission technique in which the frequency of the transmitted signal is varied (typically by a few tens or hundreds of kilohertz) in accordance with the intelligence (generally voice or music) being transmitted.

FOIRL. Fiber Optic Inter Repeater Link.

Foreground. A computer operation that takes place with visible action on the monitor, and under continuous control from the keyboard. See *Background.*

FORTRAN. FORmula TRANslator. A popular compiler language designed in the 1950s for science and mathematics applications.

FPGA. Field Programmable Gate Array.

FPS. Frames Per Second.

FPU, *computers.* Floating Point Unit.

Frame, 1. *television*. A single complete video image. In the USA, formed 30 times per second by two interlaced *fields*. 2. In data networks, the information packet and all of the preceding and succeeding signals necessary (flag bytes, preambles, frame checks, abort sequences, etc.) to convey it along the data link.

Freeware, *computers*. Software made available without charge.

Friendly, *computers*. A program or operating system that is particularly easy for the operator to learn.

FSK. Frequency-Shift Keying. Digital data transmission by changing signal frequency to represent binary states.

FTP, *internet*. File Transfer Protocol. The common method of transferring files via internet.

Full adder, *computers*. A binary adder capable of handling a carry input; as opposed to a *half adder*, which cannot handle a carry input.

Full-duplex. A communication system between two entities in which either entity can transmit simultaneously.

Fuse. A protective device that opens a circuit upon overcurrent.

Fuze. A device used to detonate an explosive charge.

FWIW, *internet*. For What It's Worth.

Gate TurnOff switch, GTO. A switching device similar to an SCR, but able to be turned off from its gate terminal.

Gaussian distribution. A continuous symmetrical distribution of data about the mean; normal distribution; bell curve.

Gender. Referring to the male (plug) or female (receptacle) status of a connector.

GIGO, *computers*. An oft-repeated truism, "Garbage In, Garbage Out."

Glitch. A noise spike. *Informal*. A problem that causes an interruption in service or progress.

Gopher, *internet*. A net-search utility.

GPIB. General Purpose Interface Bus. The IEEE-488 standard parallel interface bus, also called the *h-p bus*.

GPS. Global Positioning System. A satellite-based navigational system using microwave radio, capable of locating a position to within about 100 feet

Greeking, *computers.* A technique for reducing a line of alphanumeric display to an illegible line of hash marks. Used in page-layout editing either to save memory or to reduce display-setup time when laying out a page.

Growler. A magnetic test device that makes a growling noise to identify shorts in the armature of a motor or generator.

GUI, *computer.* Graphical User Interface.

Gunn diode. A diode that produces gigahertz oscillations when biased at the proper voltage.

h-parameters. A set of four ratios characterizing the behavior of a four-terminal network. Termed *hybrid* because the ratios are not all of the same quantities.

Hacker, *computers.* A person who examines and/or modifies the technical details of computer programs or networks. Used by some to denote a person who does so with malicious or illegal intent.

Half adder. A logic circuit that adds two binary digits, producing the sum and carry outputs, but is unable to handle a carry input. Two half adders can be connected to form a full adder that can handle a carry input.

Hall effect. Development of a voltage across a metal or semiconductor block placed in a magnetic field.

Handshaking. Data transfer in which the transmitter sends a *request to send* and the receiver acknowledges with a *clear to send* via separate lines before data is sent.

Hard copy, *computers.* Paper printout, as opposed to a video-screen display.

Hard disk, *computers.* A magnetic bulk-storage device that usually remains with the computer and is not transported.

Hardware, *computers.* The physical computer and associated machines.

Hard-wired. Inherent in the electronic circuit connections, as opposed to *programmable.* Unalterable.

Harmonic. A sinusoid having a frequency that is an integral multiple of the fundamental frequency.

Hartley oscillator. An oscillator characterized by a tapped coil in the tuned circuit.

HDTV. High-Definition TeleVision.

Header. A strip of pins on a circuit board that may be connected to a receptacle for interconnections.

Hermetic seal. A seal preventing the passage of air, water vapor, or other gases.

Hertz. The international unit of frequency; one cycle per second.

Hertz antenna. A half-wave center-fed dipole.

Heterodyne. To mix two ac signals of frequencies f_1 and f_2 in a nonlinear device, producing two additional output frequencies, $(f_1 + f_2)$ and $(f_1 - f_2)$.

Hexadecimal. The number system in base 16, counted 0, 1, 2, 3, 4, 5, 6, 7, 8, 9, A, B, C, D, E, F. Often used in computing, where two "hex" digits represent one byte.

HF. High Frequency. The radio spectrum from 3 MHz to 30 MHz.

High-level language, *computers*.   Compilers and interpreters, as opposed to assemblers.

Hole.   A vacancy in the electron structure of a semiconductor; regarded as a mobile positive charge carrier.

Holography. Three-dimensional photography using laser light. The effect of viewing a hologram is as if one were looking through a window at a three-dimensional object.

Hooks, *computers*.   Points in a program at which it accesses RAM for an address or an instruction. Used by hackers to gain control of the program and modify or examine its function.

Host, *computers*.   A computer system that runs or assembles programs for, or otherwise provides services for another computer, called the *target* system.

HTML, *internet*.   HyperText Markup Language.   The language used to create hyperlinks on the WWW.

HTTP, *internet*.   HyperText Transfer Protocol.   The common method of data transfer on the WWW.

Hub. A term used to describe multi-port network repeaters, usually of twisted-pair networks such as 10-Base-T or LocalTalk.

HV. High Voltage.

HVAC. Heating, Ventilation, and Air Conditioning.

Hybrid integrated circuit. A combination of monolithic integrated circuits, with small, often unencapsulated, discrete components.

Hyperlink, *internet.* A picture, icon, or text phrase that automatically takes the user to a new page of data about the particular topic represented, by a point-and-click of the mouse.

Hypertext, *internet.* A text phrase serving as a hyperlink, usually identified by being underlined.

Hysteresis. A snap-action effect similar to mechanical friction. Sometimes expressed by the graphic words *slop, stickiness,* or *dead-zone.*

ICE, *computers.* In-Circuit Emulator. An emulator that plugs into the target system, allowing gradual development of new systems by incremental transitions of functions from the emulator to the target system.

Icon, *computers.* A graphic image on a display screen suggestive of the function that will be performed by selecting it, such as a hand holding a pencil for a word-processing program.

I.D. Inside Diameter.

IDE, *computers.* Integrated Drive Electronics. A hard-disk drive controller.

IEC. International Electrotechnical Commission.

IEEE. Institute of Electrical and Electronics Engineers.

IETF. Internet Engineering Task Force.

IFF. Identification: Friend or Foe.

IGBT. Insulated Gate Bijunction Transistor. A power semiconductor with an FET input for high input impedance and a bipolar output for high current with low saturation voltage.

Ignitron. A high-power controlled rectifier using a pool of mercury to emit ions for conduction.

I²L. Integrated Injection Logic. A direct-coupled bipolar-transistor logic family.

Image. A spurious signal encountered in superheterodyne receivers at a frequency $f_r + 2f_{if}$, where f_r is the received frequency, f_{if} is the intermediate frequency, and the local oscillator frequency is above f_r.

IMD. InterModulation Distortion. See *intermodulation.*

IMHO, *internet.* In My Humble Opinion.

Immediate addressing, *computers.* An addressing mode in which the operand data appears in the program list at an address immediately following the instruction code.

Impedance. The vector combination of resistance and reactance. Quantity symbol: Z. Units of *Ohms* (Ω.)

Implied addressing, *computers.* Instructions for which no operand address need be specified.

Indexed addressing, *computers.* An addressing mode in which the operand is located at an address obtained by adding a fixed value to a variable value. One of these values is commonly stored in memory and the other is contained in an internal machine register. Generally used to step through a list of data on successive passes through a program loop.

Indirect addressing, *computers.* An addressing mode in which a *pointer address* follows the op code in the program. The operand data is found at the address pointed to. Used to permit modifying the operand address without modifying the program.

Input/Output (I/O). Related to the problem of getting data into or out of a computer.

Instruction. A set of binary digits interpreted by the computer to activate certain machine functions, such as add, shift, store, etc.

Instruction cycle. The set of machine cycles required to execute a given instruction. Typically two to twenty machine cycles.

Instrument transformer. A transformer that passes voltage, current, or phase information from a high-level primary to a low-level secondary for purpose of measurement.

Integrated. A multitude of parts brought together and made one.

Interface. The circuitry or connections between a computer and an I/O device, such as a thermistor or a printer.

Interlace, *television.* To position lines made on one scan down the field in between lines made on another pass down the same field, in order to increase resolution.

Intermediate frequency. The difference frequency $f_o - f_r$ produced in a superheterodyne receiver, where f_o is the local-oscillator frequency, and f_r is the received frequency. Standard IFs are 455 kHz for AM radio, 10.7 MHz for FM broadcast, and 45.75 MHz for TV.

Intermodulation. Distortion consisting of new frequencies produced by heterodyning of two desired frequencies in an amplifier that is not perfectly linear.

Internet. The worldwide system of linked networks that is capable of exchanging mail and data through a common addressing and naming system based on TCP/IP protocols.

Internic. The organization that coordinates unique naming of all computers on the internet, using the Domain Name System.

Interpolate. A process of "reading between the lines" in a data table. Thus, if the recorded temperature was 76° at 3:00, and 80° at 5:00, we might *interpolate* that it was 78° at 4:00.

Interpreter. A program that executes compiler-type instructions immediately, one by one, instead of assembling all statements before batch execution.

Interrupt. The process of halting execution of a main program, storing intermediate results, executing an *interrupt routine*, restoring the conditions prevailing before the interrupt, and resuming execution of the main program.

Intrinsic semiconductor. Essentially pure semiconductor material, whose properties are not determined by impurities.

Inverter, 1. A logical NOT gate whose output is the opposite binary state from its input. **2.** A device for converting dc to ac by switching the dc to alternating polarities.

I/O. Input/output.

Ion. A charged particle, consisting of an electron, or an atom or molecule that has acquired a charge by gaining or losing one or more electrons.

IP. Internet Protocol.

IPX. Internetwork Packet Exchange.

IRC, *internet.* Internet Relay Chat. Real-time on-line communication via internet.

ISA. Industry Standard Architecture. See EISA.

ISAPI, *internet.* Internet Services Application Programming Interface.

ISDN. Integrated Services Digital Network.

IS-IS. Intermediate System to Intermediate System. A dynamic routing protocol for IP.

ISO. International Standards Organization.

Isolation transformer. A one-to-one transformer used to isolate equipment at the secondary from ground or from a common voltage which may exist at the primary.

Isotopes. Atoms having the same number of protons but a differing number of neutrons.

Isotropic. Having identical properties in all directions.

ISP. Internet Service Provider.

Iterative process. A mathematical technique for calculating a desired result by successively closer approximations.

ITU. International Telecommunications Union. An agency of the United Nations.

Jack. A receptacle connector into which a plug may be inserted.

Java, *internet.* A language based upon C++ which allows programmers to write one program code for all platforms of computers on the Web.

JEDEC. Joint Electron Device Engineering Council.

JIT. Just In Time. A production management plan under which all components for a product are delivered just in time to be incorporated into the finished unit, thus eliminating the inventory storage problem.

JK flip flop. A triggered flip flop capable of being set, cleared, toggled, or unchanged in response to a trigger signal, depending upon the states of two inputs, J and K.

K, *computer.* Kilo, meaning 2^{10} or 1024. Not to be confused with the metric prefix "k" meaning 1000.

Kerning, *word processing.* The art of placing letters (especially large-size letters) closer together to achieve a balanced look. For example, the o might be moved under the right top arm of the T in Tot to make the white space between the letters appear more equal.

Kludge. A word used to describe a solution to a problem that lacks elegance.

Klystron. A UHF electron tube in which an electron beam is bunched by electric fields and fed into a resonant cavity.

Kurtosis. The degree to which a spectrum of frequencies is centered about a mean.

Lagging. Occurring later in time.

LAN. Local Area Network.

Landscape. A display or printout that is wider than it is long. Opposed to *portrait*.

Laplace transform. A technique for solving differential equations by reducing them to algebraic equations.

Laser. Light Amplification by Stimulated Emission of Radiation. An electron device that produces a beam of coherent light.

Latch. A storage element that retains digital data while the data source is in a state of change.

Latency. In data transmission, the delay that occurs while information remains in a device's buffered memory before it can be sent.

Lattice. A pattern of positions on a regular grid of lines. Descriptive of atomic structure in a semiconductor.

LCD. Liquid Crystal Display. A light-reflective alphanumeric and/or graphical display characterized by low current consumption.

Lead time. The time required for a manufacturer to develop a product and tool up to be ready to produce it at a given time.

LeClanche cell. The common carbon-zinc dry cell.

LED. Light Emitting Diode. A semiconductor that glows red, yellow, green, or blue when excited by typically 1.5 V, 15 mA, dc.

LIFO. Last In, First Out. A data stacking and retrieval scheme.

Light pipe. A transparent plastic rod that transmits light from one end to the other.

Linear structured programs, *computers.* Programs in which all instructions follow in program-counter sequence, except for subroutine calls, iterations of a block of instructions, or bypassing of a block of instructions.

LISP. A computer language.

Lissajous figure. (Say *lee-sah-jhoo*; no accent on any syllable.) The pattern on an oscilloscope when two sine waves of related frequencies are applied to the vertical and horizontal deflection systems, respectively. The figure may be a line, an ellipse, a figure-eight, or a more complex series of loops.

Listserver, *internet.* A utility that allows members of a group to have e-mail messages sent to all members of the group with one command.

Literal. A letter or group of letters representing a constant value.

Load. *noun.* The device that receives the output of a signal source. *verb.* To cause a drop in output voltage by connecting a load or by decreasing its resistance.

LOC. Lead On Chip. An integrated circuit manufacturing process in which the leads are a part of the chip itself, rather than part of a larger external package.

Logic function. An expression of the relationship between binary inputs and output of a circuit; including AND, OR, NOT, NAND, NOR, and XOR (exclusive OR.)

LOGO. A graphics oriented computer language, often used to introduce children to computing.

Long tail. An emitter follower or common-emitter amplifier using an emitter current source consisting of a high-value resistor fed from a relatively high-voltage supply.

Look-ahead carry. Use of separate logic to generate the carry output in adders and counters. Eliminates waiting for the count to ripple through all bits and permits fast cascading of units.

Lookup table. A computer technique in which a required function (say, temperature) is obtained from data stored in a data list and fetched out in response to an input variable (say, binary count from an ADC driven by a thermistor).

Loop, *computers.* A series of instructions executed a number of times repetitively.

Loran. LOng Range Aid to Navigation. A system using pulse transmissions from widely spaced stations and operating at a frequency of 100 kHz.

LSB, 1. Lower SideBand. 2. Least-Significant Bit.
   3. Least-Significant Byte.

LSD. Least-Significant Digit.

LSI. Large Scale Integration. Semiconductor ICs having over 100 gates or equivalent.

Lurking, *internet.* Reading news in newsgroups for some time before posting to the group.

LVDT. Linear Variable Differential Transformer. A position-measuring device with a typical range of a few centimeters, and using ac excitation.

Machine cycle, *computers.* One complete fetch/execute cycle. May consist of one or several clock cycles.

Machine language. A program in a form immediately usable by the computer, usually consisting entirely of binary, hexadecimal, or octal digits.

Macro, 1. *general.* Large in size or scale; opposed to *micro.* **2.** *computers.* A number of instructions or keyboard steps grouped together and accessed by a single command.

Magnetostriction. A change in length of a magnetic rod when placed in a magnetic field; Joule effect.

Magnetron. A UHF power-oscillator electron tube using magnetic fields and cavity resonators.

Mail merge. A computer word-processing function that allows the address, greeting, and other specifics of a letter being mailed to a group of addressees to be taken from a data base, so each letter appears personalized.

Mainframe. A full-sized computer, generally costing $1 000 000 or more and requiring a special room and staff for its operation.

Mains (*British*). The ac power lines.

Majority carriers. Electrons in N-type material and holes in P-type material.

Male connector. One in which the main or center conductor is a plug that fits into a receptacle.

MAP. Manufacturing Automation Protocol. A standard for communication between intelligent machines in a manufacturing environment.

Marconi antenna. A quarter wave fed against ground.

Maser. Microwave Amplification by Stimulated Emission of Radiation. A low-noise amplifier similar in principle to the laser.

Masking, 1. Coating selected areas of a semiconductor wafer, leaving other areas exposed for diffusion, etching, or metallization. **2.** *computers.* Preventing selected bits from being seen or acted upon by the computer.

Master-slave flip flop. A pair of flip flops treated as a unit. The master changes state on the rising clock pulse and the slave follows on the falling clock pulse. This removes ambiguity where the clock falls during an input signal transition.

Maximum Usable Frequency (MUF). The highest radio frequency that will be reflected by the ionosphere at a given time.

MDA. Monochrome Display Array. A video standard for personal computers, with a resolution of 720×348 pixels.

Media, *computers*. The physical environment in which the data is carried.

Memory map. A vertically oriented chart showing the uses assigned to various memory address-ranges in a particular computer system.

Memory-mapped. A computer architecture that accesses I/O devices as if they were memory locations.

Mesa transistor. A precursor of the planar transistor in which a wafer is etched down around protruding base and emitter regions.

MF, 1. Medium Frequency. The radio spectrum from 300 kHz to 3 MHz. **2.** Metal Film; a resistor structure.

Mho, *obsolete*. Siemens; reciprocal of Ohms.

Microchannel architecture. An internal bus proprietary with IBM personal computers.

Microcomputer, 1. A fully operational computer built around a microprocessor IC; generally costing $20 000 or less and used by one individual at a time. **2.** An integrated circuit containing a microprocessor, program and data memory, and input/output circuitry in one package.

Microprogramming. A series of instructions hard-wired into a computer's MPU to permit it to perform many fetch-execute cycles in response to one instruction. Permits relatively complex instructions, such as *divide* to be included in a microprocessor's instruction set.

MIDI. Musical Instrument Digital Interface. A standard for computer controlled or generated music systems.

Miller effect. A decrease in input impedance of an inverting amplifier caused by amplified voltage appearing across an impedance connected from output to input.

Miller integrator. A linear ramp generator that operates by charging a capacitor connected from output to input of an inverting amplifier.

MIME, *internet*. Multi-purpose Internet Mail Extensions. A system that lets you send computer files as e-mail.

Minicomputer. A computer midway in size and capability between a mainframe and a microcomputer, typically costing $50 000 to $500 000 and occupying a corner or side of a room.

MIPS. Million Instructions per Second. A measure of computing speed.

MIS. Management of Information Systems

Mixer, **1**. A linear device used to combine several audio or video signals in any desired proportion. **2**. A nonlinear device used to heterodyne two signals.

Mnemonics, *computers*. A short series of letters suggestive of a complete instruction name, and used in writing source code. Examples: ROR for *rotate right*, and JSR for *jump to subroutine*.

Modem. MOdulator/DEModulator. A device that interfaces a computer to a telephone line, usually by frequency modulation of an audio tone.

Modulation index. In FM radio, the deviation in carrier frequency divided by the modulating frequency.

Modulator. A device that varies the amplitude, frequency, or phase of an ac signal for purposes of transmitting information.

Monitor, *computers*. **1**. A housekeeping program that permits a computer to communicate with its I/O devices, step through addresses, check register conditions, and so on. **2**. A two-dimensional display device for a computer; usually CRT-based.

Monolithic integrated circuit. An IC fabricated from a single chip of semiconductor material, usually silicon.

Monotonic. A characteristic of digital-to-analog converters guaranteeing that an increasing input count will in no case produce a decreasing output voltage.

Monte Carlo analysis. A mathematical technique that randomly changes component values over their tolerance ranges to determine the overall effect on the system.

MOS. Metal Oxide Semiconductor. A technology for producing transistors and ICs using field-effect transistors having their gates isolated from the channel by an oxide of silicon.

MOSFET. Metal Oxide Silicon Field Effect Transistor.

Motherboard. A circuit board containing a number of sockets designed to hold other printed-circuit boards (the *daughter* boards.)

Mouse, *computers*. A hand-held device with one or more buttons that is moved about on a desk or pad to position a pointer or cursor on the display.

MOV. Metal Oxide Varistor. A two-terminal semiconductor surge-voltage limiting device.

MPU. Main Processing Unit of a computer.

MRI. Magnetic Resonance Imaging. Production of images similar to X-ray photos, but obtained through nuclear magnetic resonance techniques.

MSB. Most-Significant Bit; Most-Significant Byte.

MSD. Most-Significant Digit.

MSI. Medium Scale Integration. Semiconductor integrated circuits having over ten gates or the equivalent.

MTBF. Mean Time Between Failure.

Multiplex (MUX). Any of several techniques for sending multiple data signals over a single wire or radio link.

Multitasking, *computers*. An operating system feature in which two or more functions can be undertaken apparently simultaneously. Most commonly used simply to allow two application programs, for example, a word processor and a graphics program, to be up and running and instantly available to the user.

Narrow-band FM (NBFM) . Frequency modulation with deviations less than ±15 kHz.

Native. Something that is a standard part of a computer system, such as an operating system, a language, or a protocol.

NBS. National Bureau of Standards.

NC. **1.** Normally Closed (switch contact.) **2.** Numerical Control. **3.** No Connection.

NEC. National Electric Code.

Negative resistance. A property of tetrode vacuum tubes and tunnel diodes in certain voltage ranges wherein increased voltage causes decreased current.

NEMA. National Electrical Manufacturer's Association.

Nesting, *computers*. A programming technique of putting one loop inside another loop.

Netiquette, *internet*. Etiquette; acceptable behavior.

Neutralization. A technique for combating self-oscillation in an amplifier by providing negative feedback to cancel positive feedback of stray capacitance.

Newsgroups, *internet*. A system similar to bulletin boards, by means of which users can post messages and reply to messages already posted.

NF. Noise figure.

Nibble. A unit of binary data consisting of four bits.

NMR. Nuclear Magnetic Resonance. See MRI.

Node. A networked computing device that can initiate and respond to communication from other networked devices that employ similar protocols.

Nondestructive readout. Reading memory data without erasing that data.

Non-Volatile. Information that will remain usable by a computer despite loss of power.

Normalize. To multiply all data by a factor such that the reference datum level is unity (one).

NOVRAM. NOnVolatile RAM. A semiconductor memory that can be written to or read equally as fast, and that does not lose data when power is removed. Usually effected by integration of a battery and a static RAM.

NRZ. Non-Return to Zero. A data transmission or recording scheme in which binary bits are represented by level *changes*, rather than by the levels themselves. Thus the level does not return to zero after each bit is sent.

NT-1. Network Termination 1. A common interface between the ISDN and user equipment.

NTSC. National Television Systems Committee. The designation for the analog TV standard used in the USA.

Null modem. A connector that links a computer's serial output line back to its serial input. Used for testing.

Null set. A set or group with no members in it.

Numerical Control. A term applied to machine tools to indicate automatic digital control.

Object program, *computers*. Program in a form usable by the computer, as opposed to *source code*. Machine code.

OCR. Optical Character Reader; Optical Character Recognition.

Octal. The number system in base eight, counted Ø, 1, 2, 3, 4, 5, 6, 7, 11, 12, 13 ...

O.D. Outside Diameter.

OEM. Original Equipment Manufacturer.

Off-line. Computer operations performed on data that was gathered at an earlier time.

On-line. Computer or control operations that are, and must be, done as the process being controlled is underway.

Operand, *computers*. The data upon which an instruction is to operate.

Operating point. The combination of voltages and currents at which a transistor or vacuum tube is biased.

Orthogonal, *computers*. A term used to describe an instruction set in which all addressing modes are available for all instructions and on all registers.

OS/2. IBM's operating system for 80X86-based systems.

OSHA. Occupational Safety and Health Administration.

OTOH, *internet*. On The Other Hand.

Overlaying, *computers*. Transferring programs from bulk to high-speed storage during processing of earlier parts of the program to make maximum use of high-speed storage.

Overtone crystal. A crystal designed to oscillate on a harmonic of its natural fundamental frequency.

Overwrite. An error condition that occurs when a process stores data in a location that is not ready to accept it.

Packet. A radio transmission scheme in which discrete amounts of data are sent repeatedly until the receiving unit verifies that they have been copied correctly.

Packing, *computers*. To put two characters (often decimal digits) in one byte of data; four bits to the character.

Pad, 1. An attenuator used to stabilize impedance on a line. **2.** An area on a printed circuit board designed to accept solder. **3,** *computers*. To fill a data field with blanks.

Page, *computers*. A block of memory addressable by one machine word; for example, a 256-word block in a microcomputer having an 8-bit word length.

PAL, 1. Phase Alternate Line. A color television standard used in some countries outside the USA. **2.** Programmed Array Logic. A logic circuit in which the relationships between inputs and outputs are user programmable. Also PLA, Programmable Array Logic.

PAM. Pulse Amplitude Modulation.

Panoramic receiver. A receiver having a CRT display of signal strength vs. frequency for all signals near the received frequency.

Parameter. The measurable characteristics or variables of a component or circuit.

Parametric amplifier. A UHF amplifier whose operation is based on changing the reactance of a semiconductor or tube with a locally generated "pumping" voltage.

Parasitic element. An antenna element that reflects or reradiates energy but is not directly connected to the transmission line.

Parasitic oscillation. An undesirable high-frequency oscillation caused by stray inductance or capacitance and at a frequency unrelated to the operating frequency.

Parity. A binary digit added to a word to make the sum of all **1** bits always even or always odd. Used to catch machine errors.

Pascal. A compiler language noted for its encouragement of linear structured programming.

Passivation layer. An oxide layer deposited on the surface of a transistor or IC to preserve it from contamination.

Passive component. A component that is not capable of amplification or switching action.

PC. **1**. Personal Computer. **2**. Printed Circuit. **3**. Program Counter.

PCB. Printed Circuit Board.

PCM. Pulse Code Modulation.

PDA. Personal Digital Assistant.

Peltier effect. Heating or cooling of a junction of dissimilar metals when a current is passed through them.

PEP. Peak Envelope Power. A term used in single-sideband transmission to refer to average power at the maximum audio signal peaks.

Peripherals, *computers*. Hardware used in conjunction with the main processing unit, such as printers, data-storage devices, displays, and so on.

Phase. The time relationship of one waveform with respect to another of the same frequency, expressed in degrees or radians, with 360° representing one complete cycle.

Phase Detector. An electronic circuit that produces an output voltage proportional to the phase difference between two input signals of the same frequency.

Phase-Locked Loop (PLL). A circuit in which a voltage-controlled oscillator is kept in an exact frequency relationship with a reference signal by a control voltage generated by a phase detector.

Photoresistive. Changing resistance with light intensity.

Photovoltaic. Generating a voltage as a result of light radiation.

PID. Proportional-Integral-Derivative control. A control system in which the restoring force is proportional to the deviation from the set point, increases as small errors are accumulated over time (integral), and increases if a large rate of change from the set point is sensed (derivative.)

Piezoelectric effect. (Say *pea-AY-zoh*.) In a quartz crystal, generation of a voltage in response to a mechanical vibration, or production of a mechanical vibration in response to an applied ac voltage.

PIN diode. A semiconductor diode formed by layers of P-doped, Intrinsic, and N-doped silicon.

Pinchoff voltage. The reverse gate-source voltage that reduces channel current of an FET to a specified near-zero level.

Ping. A network diagnostic utility that sends an Echo Request to a distant node, which must return an Echo Reply packet back to the originating node.

Pink noise. Noise having more power content toward the low end of the frequency spectrum.

Pipelining, *computers*. A technique of fetching an instruction while the processor is internally completing the previous instruction. Used to improve speed.

PIV. Peak Inverse Voltage. A diode rating. Also PRV.

Pixel. Picture Element. The smallest component of a digitally generated or stored image.

PL. Private Line. See *CTCSS*.

PLA. Programmed Logic Array.

Planar process. A process of transistor and IC manufacture in which junctions are diffused into an epitaxial layer.

Platform, *computers*. A computer system, considered in its capacity to run applications programs or support peripheral devices.

PLC. Programmable Logic Controller.

PLCC. Plastic Leaded Chip Carrier. A plastic IC package with leads (as opposed to *leadless*).

PLL. See *Phase-Locked Loop*.

PMFBI, *internet*. Pardon Me For Butting In.

Point-contact. An early process for diode and transistor fabrication in which wire points are bonded to a base piece of doped semiconductor.

Pointer, *computers*. A register or memory location that contains an address to be used to direct the computer to desired data or program routines.

Polling. A means of external data access in which a device may input information only when it is requested to do so by a controller device.

Portable. Computer software that can be run on many different types of machines.

Portrait. A display or printout that is taller than it is wide.

POS. Point Of Sale.

Postscript. A software package copyrighted by Adobe systems and used to create textual and graphical images from mathematical descriptions of their contours.

Pot. Potentiometer; a resistor with a variable tap.

Potential. Electromotive force. Voltage.

Potential barrier. A semiconductor region in which charge carriers are repelled and may be turned back unless the external applied voltage is sufficient to overcome them.

Potentiometer, 1. An instrument for zero-current measurement of dc voltages by balancing an internal source voltage against the voltage to be measured. **2.** A three-terminal adjustable resistor.

Potting. A rubber or plastic insulating compound in which an assembly may be encapsulated for protection from vibration, moisture, etc.

Power factor. The ratio of true power to apparent power in an ac circuit.

PPC. Process-to-Process Communication.

PPI, 1. *radar.* Plan Position Indicator. A display screen representing the scanned area much as does a map.
   2. *computers.* Programmable Peripheral Interface. A microprocessor parallel interface chip by Intel Corp.

PPM. Part Per Million. Equal to 0.0001%.

PPP, *internet.* Point to Point Protocol. A dial-up internet connection.

PRF. Pulse Repetition Frequency.

Program, *computers.* A sequence of instructions to be followed by a computer.

Programmer's model, *computers.* A chart representing all the registers of a computer that are accessible to the programmer by name and bit length.

PROM. Programmable Read Only Memory. A semiconductor memory in which writing of data is limited by requiring special voltages, longer time, etc.

Prompt, *computers.* A character, such as a colon (:) or arrow (→), which is generated on a monitor by a computer program, and indicates that the user is to enter data from the keyboard.

Propagation time. The time required for a signal to reach one point from another, be it through free space, a transmission line, an amplifier, or a logic gate.

Proprietary. Information or technology owned by an individual or a company.

Protocol, *computers.* A set of rules agreed upon to allow two computers to communicate with one another.

Pseudocode, *computers.* A logical series of statements describing in plain English the various steps that comprise a program. Used as an alternative to flowcharting because it is easier to generate on a word processor, and because it encourages the writing of linear programs.

PTO. Permeability Tuned Oscillator. An oscillator whose frequency is varied by inserting or removing a magnetic core in the inductance.

PTT. Push To Talk. A common microphone button.

Puck, *computers.* A mouse-like device with crosshairs for more accurate positioning over a *tablet.* Used in *CAD.*

QA. Quality Assurance. Also *QC;* Quality control.

Quadrature. A 90° phase relationship.

Queue, *computers.* A waiting line.

Quiescent. At rest; bias condition without signal.

Quote marks, *internet.* Usually a (>) sign before each line of text to indicate material reproduced from a previous newsgroup posting or e-mail.

QWERTY. The standard typewriter keyboard arrangement.

Race. An ambiguous condition wherein two logic levels are changing states simultaneously, the result depending on which one changes first.

Radar. RAdio Detection And Ranging.

Radial leads. Component leads situated at the end of the component, allowing vertical mounting. See *Axial.*

Radix. The base of a number system.

RAM. Random Access Memory. A memory in which each word can be read as quickly as any other word. Commonly used to refer to a *read/write* memory, as opposed to a *read only* or a *read mostly* memory.

Random logic. A logic system composed of hardware logic gates (NAND, NOR, etc); as opposed to *programmed* logic.

Raster. The pattern of scanning lines covering the CRT face in a television receiver.

Rat's nest. A disparaging name for an *ad hoc* computer program.

R & D. Research and Development.

Reactance. The ratio of voltage divided by current for a circuit element, when the voltage and current are ac sine waves out of phase by 90°. A property of capacitors and inductors that allows them to store energy from an ac source on one quarter cycle and return it to the source on the next quarter cycle.

Read. To pick up data from a register, bus, or memory.

Real estate. Area on a printed circuit board or IC chip available for placement of components.

Real-time operation, *computers.* Processing of input data as it is received, and producing outputs that affect the process which is supplying those inputs; as opposed to *batch processing.*

Recursive routine, *computers.* A subroutine that can call itself without corruption of data or tying the processor up in an endless loop.

Re-entrant routine, *computers.* A subroutine that can be interrupted and called again by the interrupting routine without corruption of data.

Reflectometer. A directional coupler used to measure reflected waves or standing waves on a transmission line.

Refresh, *computers.* The process of recharging the storage capacitors in a dynamic RAM.

Register, *computers.* A binary storage device, generally for one word of data, and usually capable of logic or arithmetic operations.

Register-oriented, *computers.* A computer architecture in which memory is accessed relatively infrequently and most operations are performed in a number of *working registers* within the MPU.

Rel. Reliability.

Relative addressing, *computers.* A addressing mode that references a memory location in terms of its distance from the current program-counter contents, rather than in absolute terms.

Relaxation oscillator. An oscillator whose frequency is determined by the charging time of an RC circuit.

Relay rack. A frame to accommodate standard 19-inch-wide panels for equipment mounting.

Repeater. A device that receives weak signals and transmits corresponding stronger signals, sometimes on a different frequency and sometimes after signal processing.

Resistance. The ratio of voltage divided by current for a circuit element, when the voltage and current are either dc or ac in-phase.

Resonance. The condition in which the frequency of an externally applied force equals the natural oscillation frequency of a system.

Reverse engineering. Disassembling or examining a competitor's product with intent to incorporate its features into one's own product.

RFC. Radio Frequency Choke. An inductor, typically 0.1 mH to 10 mH.

RFD, *internet*. Request For Discussion, to begin, alter, or delete a newsgroup.

RFI. Radio-Frequency Interference.

RGB. Red-Green-Blue. A system of transmitting color television signals on three lines by their separate colors; as opposed to *composite video*.

Rheostat. A two-terminal variable resistor, especially one carrying high current.

Ringing. A damped oscillation following a step change in input.

RIP. Routing Information Protocol.

RISC. Reduced Instruction-Set Computer. A computer designed for very fast execution of a limited number of instructions.

Risetime. The time required for a voltage to rise from 10% to 90% of its full value, when a much faster forcing function drives it from 0 to 100%.

RIT. Receiver-Incremental Tuning. A control in a transceiver that allows the receiver to be tuned to a slightly different frequency than the transmitter.

rms. The Root of the Mean after the Square. A mathematical technique of calculus used to determine the dc equivalent of an ac waveform in terms of its power-delivering capability. For sinusoids, it yields the familiar fact that $V_{rms} = 0.707\ V_{pk}$.

Robot. A multipurpose manipulative device, generally used in a manufacturing environment to achieve better reliability, economy, or safety than is possible with human operators.

ROM. Read Only Memory. A semiconductor memory into which data has been permanently stored. Often used loosely to refer to nonvolatile memory that cannot be overwritten by the target computer.

ROTFL, *internet*. Roll On The Floor Laughing.

Router. A device that forwards packets between networks.

RS-232C. A serial data-transmission standard using voltages below –5 for logic **1** and above +5 for logic **Ø**.

RTD. Resistive Temperature Device. A two-terminal component whose resistance varies with temperature.

RTTY. Radio TeleTYpe.

SAA. Systems Application Architecture.

Salient pole. In a motor or generator, a magnetic pole that protrudes from, rather than blends into, the cylindrical shape of rotation.

Sampling oscilloscope. An oscilloscope capable of displaying repetitive signals well above 1 GHz by fast switching of diodes to obtain samples of the signal at progressively delayed points on the wave.

SAP. Service Access Point. The interface between one protocol and another.

SASE. Self-Addressed Stamped Envelope.

Saturation. The point at which increasing one quantity no longer has an effect on a second quantity. Commonly applied to base current vs. collector current of a transistor, input vs. output voltage of an amplifier, and magnetizing current vs. magnetic flux.

SAW filter. Surface Acoustic Wave filter. A ceramic device that behaves as a tuned circuit at a fixed frequency.

SBS. Silicon Bilateral Switch.

SCA. Subsidiary Communications Authorization. Subscription music service transmitted by subcarrier from FM broadcast stations.

SCR. Silicon Controlled Rectifier. A three-terminal semiconductor that switches into conduction from anode to cathode when a specified current is applied to the gate.

Scratch pad, *computers*. A small, fast-access RAM used for temporary data storage and retrieval.

Script. A set of instructions that a computer can execute.

SCSI. (say *SCUZ-ee*). Small Computer Systems Interface. A fast serial data standard for desktop computer interfacing to hard disk drives, laser printers, etc.

SDLC. Synchronous Data Link Control.

SECAAM. A color television standard used in some countries outside the USA.

Second detector. The demodulator that converts the IF signal to AF in a superheterodyne receiver. The "first detector" is usually not so called, but is termed instead a *mixer* or a *converter*.

Seebeck effect. Voltage generated across a junction of two dissimilar metals, which is a function of temperature.

Serial. Handling data sequentially rather than simultaneously.

Server, *internet*. A central computer that supplies information on demand to other computers on the network.

Servo system. A feedback control system in which the output response is a mechanical position.

Set point. The value of a controlled variable to be maintained by a process controller; demand point.

Shaded pole. A starting scheme for ac induction motors in which the stationary poles are split, with a heavy copper ring around one portion to retard magnetic field on that side of the pole.

Shadow RAM. A semiconductor memory consisting of a volatile RAM and a backup EAROM. Used to achieve nonvolatility and fast writability in one device.

Shareware, *computers*. Application programs that users are allowed to obtain and evaluate without charge. The user is expected to send payment for the software if he or she intends to use it on a regular basis.

Shell, *computers*. An application program that serves as an interface between the user and a more difficult to use operating system.

Shell account, *internet*. An early internet connection scheme, basically a text-only mode. With the advent of graphics-based internet services such as WWW and browsers such as Netscape, shell accounts have become all but obsolete.

SHF. Super High Frequencies. 3 GHz to 30 GHz.

Short circuit. An undesired or temporary low-resistance path.

Shout, *internet.* To use ALL CAPITALS on the net. Considered rude, unless used sparingly.

SI. System International. The complete set of standard quantities and units in the metric system.

Sidebands. A span of frequencies above and below a modulated carrier, produced when modulation distorts the perfect sine-wave shape.

Signature analysis. A digital troubleshooting technique in which a serial data stream at a test point is compared with the expected data stream. Requires extensive setup to determine the expected data streams under various conditions.

Significant figure. A part of a number whose true value is known, and is not simply the result of mathematical computations or of zeros serving as place holders. Significant digit.

SIMM. Single-In-line Memory Module. A plug-in memory module having a single row of contacts.

Simplex operation. Radio communication requiring manual or automatic switching between talk and listen periods.

Single-ended. Having the signal appearing from one line to ground, rather than differentially between two lines balanced around ground.

Single sideband (SSB). A form of amplitude modulation in which the carrier and one sideband are removed and all transmitter power is concentrated in the other sideband. Its advantages include: half the bandwidth, freedom from heterodyne howl with other carriers, and better immunity from fading. Disadvantages include: increased circuit complexity, reduced fidelity, and critical receiver tuning.

Sinusoid. A signal having the shape of a sine wave, without regard for its phase.

SIP. Single In-line Package. An IC or small circuit designed with a single row of pins to be plugged into a socket.

Slewing. Moving as rapidly as possible from one point to another.

SLIP. Serial Line Internet Protocol. A dial-up protocol for carrying IP information over serial links.

SMD. Surface-Mount Device. ICs and other components designed to be mounted and soldered on a circuit board without leads protruding through holes in the board.

SMDS. Switched Multimegabit Data Service.
A metropolitan area packet switching data network using T-1 and T-3 lines.

SMT. Surface-Mount Technology.

SMTP. Simple Mail Transfer Protocol.

Snap-action diode. A diode that transitions abruptly from conduction to nonconduction.

SNMP. Simple Network Management Protocol. A de facto standard for management of networked devices.

Software. Computer programs.

SOIC. Small Outline Integrated Circuit.

Solid-state device. A device that controls electric current within solid materials, as opposed to vacuum, gases, or liquids.

Sonar. SOund Navigation And Ranging.

SONET. Synchronous Optical Network.

SOS. Silicon On Sapphire. A semiconductor production technique.

SOT. Small Outline Transistor.

Source program, *computers*. User-readable statements before they have been converted into machine language.

Spaghetti code. See *Ad hoc*.

Spam, *internet*. Unwanted messages, generally sent in bulk for solicitation or proselytizing.

SPARC. Scalable Processor Architecture. A *RISC* computer system from Sun Microsystems.

SPICE. Simulation Program for Integrated-Circuit Engineering. A popular family of computer-aided circuit-analysis programs.

Spooling, *computers*. Sending a series of packets of data from a computer to its destination.

Spreadsheet. A computer applications program designed to display multiple data entries in rows and columns, and to perform calculations on these data according to user-specified rules. A teacher's grade book is a common example.

Sprite. A small computer graphic capable of being superimposed on a larger background graphic in any position without destroying the background.

Spurious signal. Unwanted signals of chance or questionable origin.

SPX. Sequential Packet Exchange.

Square-law detector. One whose output is proportional to the square of the input.

Squirrel-cage rotor. A motor armature containing copper rods connected to copper end disks, the entire assembly embedded in a relatively low-conductivity iron core.

SRAM. See *Static RAM*.

SSB. See *Single SideBand*.

SSI. Small Scale Integration. Semiconductor integrated circuits with fewer than ten gates or equivalent.

Stack, *computer*. An area of memory designated to store the contents of the machine registers upon servicing an interrupt or subroutine; and to store user data that will be retrieved by the order in which it was stored, rather than by name or absolute address location.

Stagger tuning. Adjusting a number of tuned circuits to slightly different frequencies to give a wider, flat-topped overall response curve.

Standing-Wave Ratio (SWR). The ratio of maximum to minimum voltage or current along a transmission line. Equal to unity if source, line, and load impedances are equal and purely resistive.

Star. A network topology that is constructed by connecting computing devices to a common device.

Static. 1. At rest; zero signal activity; dc. 2. Electric charges accumulated on nonconductors (as opposed to *current* electricity, which is charge in motion).
   3. Radio interference caused by static discharges in the atmosphere.

Static RAM. A semiconductor memory whose data will be preserved as long as power to the chip is maintained.

Stepper motor. A rotating magnetic actuator in which rotation is achieved in discrete steps by applying voltages to different windings of the machine, sometimes in reversing polarities.

Stochastic. Relying on trial and error or probability for a solution; opposed to *algorithmic*.

Strain gage. A resistive element designed to be attached to the surface of a mechanical-support member, and to change resistance in a predictable way when that member is subjected to stress.

String, *computers.* A contiguous file of data, usually representing text.

Strobe. To gate ON and OFF at a regular rate.

Stub. A short section of transmission line connected in parallel with the main line and used for tuning, trapping, or impedance matching.

Subcarrier. A frequency that modulates a carrier wave, and is itself modulated by another lower-frequency signal.

Subroutine, *computers.* A section of a program that performs a well-defined and frequently used function. The program counter leaves the main program, counts through the subroutine, and then returns to the main program.

Substrate. The base upon which a transistor or IC is fabricated.

Superheterodyne. A receiver in which all received signals are converted to a fixed intermediate frequency for amplification and selectivity before demodulation. Superhet.

Super VGA, SVGA. A video graphics adapter with an 800 × 600 pixel resolution in 256-color mode, or 1024 × 768 pixel resolution in 16-color mode.

Surge. A brief increase in voltage or current.

Susceptance. The negative reciprocal of reactance. Quantity symbol: B. Units of Siemens (S.)

Switched 56. A dial-up communication service that offers a fractional portion of a T1 line; also called Fractional T1.

SX. A suffix used with the intel 486 processor to denote a built-in math coprocessor.

Synchro. A device used to duplicate an angular position at a remote location via multiwire connection.

Synchronous. In time coincidence; in step.

Sync pulse. In facsimile and television, a pulse transmitted at the end of a line or field to keep the receiver in synchronism with the transmitter.

Syntax. The rules of a programming language.

Synthesize. A technique for developing a large number of frequencies from a few master oscillators by using phase-locked loops, heterodyning, and digital frequency division.

T-1. A telecommunications technology for wide area networks (WAN), with a transmission speed of 1.554 MBPS over 24 multiplexed 56 KBPS channels.

T-2. Four multiplexed T-1 lines offering a communication channel at 6.3 MBPS.

T-3. Twenty-eight multiplexed T-1 circuits with a bandwidth of 44.736 MBPS.

TA, *computers*. Terminal Adapter. A device for connecting digital to non-digital devices.

Tablet, *computers*. A rectangular pad with a pencil-like stylus attached to a computer; used to create lines and shapes, and to select commands in a drawing program.

Target system, *computers*. The system that the program under development is to run on.

TC. 1. Thermocouple. 2. Temperature Coefficient.

TCP / IP, *internet*. Transfer Control Protocol / Internet Protocol.

TDD. Telephone Device for the Deaf.

TDR. Time Domain Reflectometer. An instrument for determining characteristics of a transmission line, and for locating breaks, by sending a pulse down the line and measuring the timing and other characteristics of the reflected pulse.

Telnet, *internet*. A system for allowing access to a home computer's network capabilities from a remote computer.

Ten-base 2, *internet*. A coaxial-cable-based network feed.

Ten base T, *internet*. A network feed based on twisted-pair wires.

Terminal. 1. A metal hook, post, or eyelet to which a wire may be connected. 2. A keyboard and display screen that communicates with a remotely located computer. *Smart* terminals contain their own microprocessor and preprocess the data. *Dumb* terminals rely on the remote computer for all data manipulation.

TFT. Thin Film Transistor.

TFTP. Trivial File Transfer Protocol. A simplified version of FTP.

THD. Total Harmonic Distortion.

Thermal runaway. A condition in a transistor in which heating causes increased current, which causes further heating in a spiral ending in saturation or the destruction of the transistor.

Thermionic. Producing emission of electrons by heating.

Thick-film circuit. A circuit in which resistors, capacitors, and conductors are printed or painted onto a substrate, often by silk-screen printing.

Thin-film circuit. A circuit in which resistors, capacitors, and semiconductors are deposited onto a substrate molecule-by-molecule in an evacuated chamber.

Thread, *internet.* A continuing topic on a newsgroup or bulletin board.

Thyristor. A family of switching semiconductor devices, including SCRs, triacs, and diacs.

Toggle. Snap action from one state to another.

Token Passing. A network control method in which stations may only transmit when they are in possession of a special bit sequence (token) passed from station to station.

Top-down. A management technique for developing large computer programs. The assumed finished program is broken down into a series of modules, each with a single entry point and a single exit point. These may each be broken into submodules in a similar way. Modules may be assigned to individual programmers and developed simultaneously.

TQM. Total Quality Management.

Trackball. A ball typically 4 cm. in diameter captured in a base with its top surface exposed, which is rolled by the hand to position a pointer on a computer display.

Transducer. A device that converts energy from one form to another, especially one that converts some physical quantity to electric current or voltage for purposes of measurement or control.

Transient. A short pulse or oscillation, as opposed to a steady-state condition.

Transistor. TRANSfer resISTOR. An active semiconductor device having three electrodes.

Transparent buffer. One in which the data at the input side at the moment of chip select is connected to the data bus; as opposed to a *latched* buffer, in which data present when a special strobe signal is asserted is held for presentation on the bus when chip-select is asserted.

Transparent DMA. Direct memory access that does not slow the processor down in any way.

Transparent latch. A latch in which the data output follows the data input as long as the chip-select is asserted; as opposed to a triggered latch in which data is sampled only upon the transition of the chip-select input.

Transponder. A remote, usually mobile, transmitter that responds with data such as position or temperature, when called upon by a control transmitter.

Trap. A tuned circuit used to eliminate an undesired frequency.

Traveling-Wave Tube (TWT). A UHF electron tube in which a wave traveling along a helix interacts with an electron beam traveling down the center of the helix.

Triac. A triggered semiconductor device similar to an SCR, but which can be made to conduct in both directions.

Trimmer. A small capacitor or resistor adjustable by a screwdriver or thumbwheel for purposes of alignment.

Tri-state. Logic circuitry having three distinct output states: high voltage, low voltage, and high-impedance (floating).

TRMS. True rms. A meter that reads the actual rms value of any waveshape within its frequency range. Opposed to the usual *average-reading* meter that responds to the absolute *average* value of the wave and indicates 1.11 times this (which is rms value for sine wave only).

Trojan, *computers*. A program that is represented to be a desirable or beneficial application, but which actually produces a harmful or undesirable effect.

Troll, *internet*. A person who posts messages to a newsgroup with the intention of provoking large numbers of emotional responses.

True power. The power delivered to an ac circuit that is dissipated (as opposed to being returned to the source.)

TTL. Transistor-Transistor Logic. Also, T^2L. A popular logic family using bipolar transistors and a +5-V supply.

TTY. TeleTYpe.

Tunnel diode. A semiconductor diode exhibiting negative resistance between 0.2 and 0.4 V forward bias in typical units. Esaki diode.

Tuple. A set of two related data items.

Turnkey system. A system in which the user is relieved of all responsibilities for technical adjustments, having only to turn it on.

Turtle. A triangular graphics cursor used in the LOGO computer language.

TVM. Transistor VoltMeter. An analog meter, generally for voltage, current, and resistance, which achieves a high input impedance by means of transistors, generally FETs.

Two's complement. The result of complementing a binary number (changing all 1s to Øs and all Øs to 1s) and then adding 1. Negative numbers in binary are expressed in two's complement form, and have a most-significant digit of 1.

UART. Universal Asynchronous Receiver/Transmitter. A device that converts parallel data words from the transmitting computer to a string of serial data bits for transmission via a single line or channel, and performs the corresponding function at the receiver.

UHF. Ultra High Frequencies. The radio frequency range from 300 MHz to 3 GHz.

UNIX. A popular operating system for medium to larger computers.

UPC. Universal Product Code. The familiar bar code printed on merchandise packages and used for checkout at the point of sale.

Upload. To transfer programs or data files from a smaller computer system to a larger one.

UPS. Uninterruptable Power Supply. A device that supplies line voltage to a computer system for a limited time in the event of failure of the electric-service mains, usually by reliance on a storage battery. Used to prevent loss of data.

URLs, *internet.* Uniform Resource Locators.

USB. Upper SideBand. See Single Sideband.

Usenet, *internet.* An organization comprised of most internet news groups.

UTC. Universal Coordinated Time; formerly Greenwich Mean Time, GMT, or Zulu.

UTP. Unshielded Twisted Pair wiring.

V.32 bis A 14.4 kbaud modem standard.

V.34 A 28.8 kbaud modem standard.

Valve, *British.* Vacuum tube.

Vaporware, *computers.* A pejorative/humorous term for software advertised by a company but not available in fact.

VAR. Volt-Amp Reactive. The product of voltage times current, without regard to true power. Used in the rating of transformers and other devices supplying current to reactive loads.

Varactor. A semiconductor diode used as a capacitor, capacitance decreasing with increased reverse bias. Varicap.

VAX. A popular series of minicomputers from Digital Equipment Corp.

VCO. Voltage-Controlled Oscillator.

VCR. Video Cassette Recorder.

VDT. Video Display Terminal.

Vernier. A device, usually a mechanical gear reduction, used for making fine adjustments to a control setting.

VESA. Video Electronics Standards Association.

V/F. Voltage-to-Frequency conversion.

VGA. Video Graphics Array. A video display interface for IBM PC and compatible computers. Best text mode is 720×400 pixels. Best graphics mode is 640×480 in 16 colors, and 320×200 in 256 colors.

VHF. Very High Frequency. The radio frequency range from 30 MHz to 300 MHz.

VHS. Video Home System. A popular video-tape cassette format.

VHSIC. Very High Speed Integrated Circuit.

Via. A connection from one layer of a printed circuit board to another.

Video frequencies. A band extending from a few hertz to generally about 5 MHz, but in some cases as much as 100 MHz.

Virtual ground. Not actually grounded, but at ground potential for purposes of most calculations.

Virtual RAM. A portion of a computer's hard disk used as RAM memory, although of course it is much slower of access than real RAM.

Virtual reality, *computers.* Any of a number of computer-based technologies for stimulating the user with more sensory data than is provided by a basic animated TV screen and sound.

Virus, *computers.* A segment of code that attaches itself to other programs and spreads by reproducing on each disk or network that is accessed by the infected computer.

VLF. Very Low Frequencies. The radio frequencies from 10 kHz to 30 kHz.

VLSI. Very Large Scale Integration. Semiconductor integrated circuits having more than 1000 gates.

VME bus. A 96-pin computer interface standard.

Volatile memory. Read/write memory whose contents are lost if the power supply is interrupted.

VOM. Volt-Ohm Meter, or Volt-Ohm-Milliammeter.

VOX. Voice Operated transmitter. A system for automatically switching a radio transceiver from receive to transmit when the operator speaks into the microphone.

VRAM. Video Random Access Memory.

VTVM. Vacuum-Tube VoltMeter. A high-input-impedance analog meter using vacuum tubes.

VU. Volume Unit.

VXO. Variable Crystal Oscillator. A crystal oscillator whose frequency is variable by about 0.1% by means of a variable inductor or capacitor.

WAIS , *internet.* Wide Area Information Server.

Wall wart. A slang term for a small step-down transformer with integrated ac wall plug; sometimes with internal rectifier and filter for supplying low-voltage dc.

WAN. Wide Area Network.

Warm boot. The process of restarting a computer without actually turning off the power.

White noise. Random noise, having equal energy at all frequencies.

WIMP interface. Windows, Icons, Mouse, and Pull-down menus.

Winchester. A popular term for early computer hard-disk drives.

Windom antenna. An off-center-fed antenna that is 1/2 wave long on the lowest frequency used, and operates on even multiples of that frequency.

Window, 1. *computers.* A rectangular area on the display containing information relating to a particular file or application program. Several windows may appear on the display at once. 2. *manufacturing.* The region or time span between two limits within which operation is possible.

Wired-OR. The technique of connecting the outputs of several open-collector logic gates together, so that a pull low from one OR another will bring the common output low.

Wire wrap. A solderless connection made by winding a bare end of solid wire around a square post that has sharp corners.

Word, *computers.* A group of binary digits treated as a unit in data-transfer operations. Common word lengths in microcomputers are 8, 16, and 32 bits. Semiconductor memories are most commonly organized with 1- or 8-bit word lengths.

Worm, *computers.* An unauthorized segment of code that reproduces by itself, rather than by attaching to another program as does a virus. Usually fatal because it continues to reproduce until it takes up all available disk space.

WORM. Write Once, Read Many. A type of one-time user-programmable read-only memory.

Write. To store data to a register or memory location.

WWW, *internet.* World Wide Web.

WYSIWYG. (say *WIZZ-ee - WIG*). What You See Is What You Get. A display on the computer monitor that exactly matches the hard-copy printout in size, type styles, color, and other respects.

X. In binary and hexadecimal notation, a character used to indicate that any digit may appear. "Don't care."

XDSL. Experimental Digital Service Line. A high-speed data line using ordinary telephone twisted-pair wire.

XGA. Extended Graphics Array. A video standard with best text mode of 1056×480 pixels, and best graphics resolution of 1024×768 in 256 colors, or 640×480 in 65 536 colors.

XIT. Transmitter Incremental Tuning.

X-Y plotter. A servo-controlled pen that draws a graph of two variables input as voltages.

Yagi antenna. A directive antenna consisting of a driven element, a reflector dipole, and several director dipoles.

Y, I, and Q. The three elements of an NTSC color TV signal as recovered from the composite video signal. *Y* is brightness, and *I* and *Q* are color modulation components that are In-phase and in Quadrature, respectively, with the color subcarrier.

Zepp antenna. A half-wave antenna fed with open-wire line at one end. Originally used by zeppelin airships.

Zero-page addressing. An addressing mode in which the memory page number of the operand is assumed to be zero.

ZIF. Zero Insertion Force. A type of IC socket designed to reduce damage to the IC pins from frequent insertions and removals.

Zulu. Greenwich Mean Time. Universal Coordinated Time (UTC).

Index